Interdisciplinary Applied Mathematics

T0235520

Springer
New York
Berlin
Heidelberg
Barcelona
Hong Kong
London
Milan
Paris
Singapore
Tokyo

Interdisciplinary Applied Mathematics

Volume 15

Editors
S. Antman **J.E. Marsden**
L. Sirovich **S. Wiggins**

Geophysics and Planetary Science

Mathematical Biology
L. Glass, J.D. Murray

Mechanics and Materials
R.V. Kohn

Systems and Control
S.S. Sastry, P.S. Krishnaprasad

Problems in engineering, computational science, and the physical and biological sciences are using increasingly sophisticated mathematical techniques. Thus, the bridge between the mathematical sciences and other disciplines is heavily traveled. The correspondingly increased dialog between the disciplines has led to the establishment of the series: *Interdisciplinary Applied Mathematics*.

The purpose of this series is to meet the current and future needs for the interaction between various science and technology areas on the one hand and mathematics on the other. This is done, firstly, by encouraging the ways that mathematics may be applied in traditional areas, as well as point towards new and innovative areas of applications; and, secondly, by encouraging other scientific disciplines to engage in a dialog with mathematicians outlining their problems to both access new methods and suggest innovative developments within mathematics itself.

The series will consist of monographs and high-level texts from researchers working on the interplay between mathematics and other fields of science and technology.

J. David Logan

Transport Modeling in Hydrogeochemical Systems

With 50 Illustrations

 Springer

J. David Logan
Department of Mathematics
and Statistics
University of Nebraska–Lincoln
Lincoln, NE 68588-0323
USA
dlogan@math.unl.edu

Editors

S. Antman
Department of Mathematics
and
Institute for Physical Science and Technology
University of Maryland
College Park, MD 20742
USA

J.E. Marsden
Control and Dynamical Systems
Mail Code 107-81
California Institute of Technology
Pasadena, CA 91125
USA

L. Sirovich
Division of
Applied Mathematics
Brown University
Providence, RI 02912
USA

S. Wiggins
Control and Dynamical Systems
Mail Code 107-81
California Institute of Technology
Pasadena, CA 91125
USA

Mathematics Subject Classification (2000): 76S05, 35Kxx, 76Rxx, 35Qxx

Library of Congress Cataloging-in-Publication Data
Logan, J. David (John David)
    Transport modeling in hydrogeochemical systems / J. David Logan.
        p. cm. — (Interdisciplinary applied mathematics)
    Includes bibliographical references (p.).

    1. Hydrogeology—Mathematical models.   I. Title.   II. Series.
GB1001.72.M35 L65 2001
551.49′01′13—dc21   2001020439

Production managed by Michael Koy; manufacturing supervised by Joe Quatela.
Photocomposed copy produced from the author's LaTeX files.
Printed and bound by Sheridan Books, Inc., Ann Arbor, MI.
Printed in the United States of America.

9 8 7 6 5 4 3 2 1

ISBN 978-1-4419-2932-7

Springer-Verlag   New York Berlin Heidelberg
A member of BertelsmannSpringer Science+Business Media GmbH

To my grandchildren—

Shelby Elizabeth Golightly,
Jackson Douglas Golightly,
Parker George Logan

Preface

The subject of this monograph lies in the joint areas of applied mathematics and hydrogeology. The goals are to introduce various mathematical techniques and ideas to applied scientists while at the same time to reveal to applied mathematicians an exciting catalog of interesting equations and examples, some of which have not undergone the rigors of mathematical analysis. Of course, there is a danger in a dual endeavor—the applied scientist may feel the mathematical models lack physical depth and the mathematician may think the mathematics is trivial. However, mathematical modeling has established itself firmly as a tool that can not only lead to greater understanding of the science, but can also be a catalyst for the advancement of science. I hope the presentation, written in the spirit of mathematical modeling, has a balance that bridges these two areas and spawns some cross-fertilization.

Notwithstanding, the reader should fully understand the idea of a mathematical model. In the world of reality we are often faced with describing and predicting the results of experiments. A mathematical model is a set of equations that encapsulates reality; it is a caricature of the real physical system that aids in our understanding of real phenomena. A good model extracts the essential features of the problem and lays out, in a simple manner, those processes and interactions that are important. By design, mathematical models should have predictive capability. In this monograph we develop transport models in hydrogeology. Obviously, subsurface phenomena are highly variable and many mechanisms superimpose to give what we observe. Our models do not always try to include every mechanism. We may try, for example, to study only diffusion and adsorption in an effort to understand the interactions between those two processes; in doing so, we may ignore other mechanisms. Therefore, the philosophy of modeling is much different from writing down equations that include every possible effect, with a plethora of parameters, and then using a computer or software package to obtain solutions numerically.

In this monograph a mathematical model generally means a set of partial differential equations (PDEs) that simply describes a physical problem. The equations are typically parabolic or hyperbolic evolution equations. The equations arise from mass balance and relate diffusive-dispersive, advective, and reactive processes involving chemical reactions. Hydrogeologically, the monograph is less about the kinematics and dynamics of groundwater flow, and more about the transport and reaction of solutes in the groundwater. There are a large number of problems examined in detail; some of these involve more advanced mathematical techniques than one usually encounters in hydrogeological treatments, and more attention is paid to the mathematical issues. In this sense the monograph leans toward the mathematics side. Therefore, this monograph is not a systematic development of the theory of transport processes, but rather a compendium of various simple models that help us understand such processes.

The prerequisites are not extensive. No knowledge of the geosciences is required—just a good physical sense and a willingness to think about physical concepts. Although not absolutely necessary, some knowledge of PDEs will be helpful, particularly PDEs associated with elementary fluid dynamics. In later portions of the monograph some knowledge of phase plane analysis is beneficial. A student who has studied post-calculus differential equations and has read an elementary PDE text can read most parts of the text.

Understanding diffusion processes is fundamental in hydrogeology, both for dispersion of solutes and for the evolution of the head. Therefore, the the book begins in Chapter 1 with a review of some of the main ideas associated with the classical diffusion, or heat, equation. These ideas include techniques for solution of the diffusion equation on both bounded and unbounded domains with a variety of boundary conditions. The behavior of solutions is discussed, including the maximum principle. Those familiar with elementary ideas in PDEs could skip this chapter.

Chapter 2 contains the core material on the transport of solutes through porous media—their dispersion, advection, and adsorption. Many examples, which form the basis of the remainder of the monograph, are worked out in detail. Techniques include eigenfunction expansions, Fourier and Laplace transformations, perturbation methods, asymptotic analysis, similarity methods, and energy methods. I have started with this material, usually involving one-dimensional flows, rather than first covering the standard notions of hydraulics and flow patterns, which are examined in Chapter 5. With this approach, the student can appreciate right away the variety of modern environmental problems involving the transport of contaminants without having to first wade through the geometries of well location or other issues concerned with the form of sub-surface velocity fields.

In Chapter 3 we discuss special types of solutions called traveling waves. These types of solutions and their stability are greatly beneficial in understanding how different physical processes interact during the evolution of the flow. The predominance of these types of solutions throughout the monograph reflects the author's own interest in them, as well as their prominence in the literature.

Chapter 4 introduces some of the ideas of filtration theory, where particles get sieved out from the medium. There is a discussion of the Herzig–Leclerc–LeGoff model as well as some refinements. The key development is the modeling of processes where the porosity is not constant.

Patterns of flow are discussed in Chapter 5. This is a standard chapter that addresses issues associated with the mechanisms that produce velocity fields and transport. Thus, we discuss Darcy's law, the Dupuis approximation, unsaturated media, and the modeling of flow fields near extraction wells. This material is very classical and can be found in any hydrogeology book. Certainly, a logical step would have been to place this material at the beginning of monograph. However, the goal was to put emphasis on some of the concepts and models in solute transport, a subject of intense environmental interest, rather than first develop the classical subject of ground water flow.

Flow through porous rocks involves many interesting phenomena relating to porosity-mineralogy changes. Chapter 6 develops some simple models useful in understanding these process. There we study both the propagation of traveling wave fronts on unbounded domains and numerical models on finite domains. The material here can serve as a lead-in to an advanced treatment like P. Ortoleva's *Geochemical Self-Organization,* where pattern formation is discussed in detail.

There are several exercises interspersed throughout the book rather than collected at the end of sections. These are meant to extend or illustrate ideas, or verify facts stated in the exposition, rather than reinforce concepts, as would be the case in an elementary textbook.

Over one hundred references are collected at the end of the book. There has not been an attempt to provide a complete bibliography, and thus many references, especially research articles, have not been included. However, the references listed should provide the reader with a stepping stone to a thorough literature search. There are a few references listed that are not cited directly in the text.

Finally, concerning notation, hydrogeology is a subject practiced by geologists, civil engineers, applied mathematicians, and others. As a result there is not a totally standard notation used in all areas. Porosity, for example, is denoted by n, θ, or ω, depending upon the source. To make adjustments easier, there is included a list of symbols in the appendix. The appendix also includes brief sections on the numerical solution of partial differential equations with programs in MATLAB and Maple. The numerical inversion of Laplace transforms is also included.

My own introduction to mathematical hydrogeology began about a decade ago, and I owe my colleagues and students a great acknowledgment for their patience in listening to my seminar talks and courses on hydrogeology, and for sharing their knowledge with me. This monograph grew out of those seminars and courses. Special thanks go to Professors Steven Cohn, Glenn Ledder, and Tom Shores of the Mathematics and Statistics Department at UNL, Professor Vitaly Zlotnik of the Geosciences Department, and Professor Michelle Homp, a former student, now at Concordia University. Bill Wolesensky, a current Ph.D. student, has read a lot of the manuscript and made many important observations leading to clarity and correctness. During most of the writing I was supported by the National Science Foundation on grant DMS 9708421. The University of Nebraska Research Council also generously supported the efforts, as well as the College of Arts and Sciences through a one-semester sabbatical.

Lincoln, Nebraska, USA   J. David Logan

Contents

Chapter 1

The Diffusion Equation

The first chapter of this monograph deals with diffusion. Diffusion processes play an important role in hydrogeological modeling. In subsurface porous structures there are really two processes that are lumped together into a diffusion term. One is ordinary molecular diffusion caused by the random collisions between molecules, and the other is dispersion, a velocity-driven process where molecules disperse by tracking through the tortuous pathways of the medium. Both processes cause molecules, say, a contaminant dissolved in water, to spread out in the medium. In this chapter we focus on one part of the diffusion-dispersion process, namely molecular diffusion.

1.1   The Diffusion Equation

Many problems in subsurface transport give rise to partial differential equations that belong to a general class called **parabolic**. The simplest example and prototype of a parabolic equation is

$$u_t = Du_{xx}, \tag{1.1}$$

which is the **diffusion equation**. In heat transfer it is called the heat equation, and in hydrogeology it is sometimes called the dispersion equation. It models, for example, the molecular diffusion of a chemical contaminant of concentration $u = u(x,t)$, dissolved in an immobile liquid. The positive constant D, having units of length-squared per unit time, is the **diffusion constant**, which measures the ability of the contaminant to diffuse through the liquid. In a porous medium, the pathways for diffusion are restricted by the presence of the solid porous fabric, and the diffusion constant is usually reduced from its value for the case of no restricted pathways. Physically, the time over which discernible concentration changes occur in a medium with length scale L, i.e., the **characteristic time** for the diffusion process, is L^2/D. Also, D is not always constant; it can depend the spatial variable if the medium is nonhomogeneous, and it could even depend on the concentration u. In this book, however, we assume D is constant.

1

In more basic terms, the diffusion equation is the conservation equation

$$u_t = -Q_x,$$

with the flux Q given by the constitutive equation

$$Q = -Du_x,$$

which is **Fick's law**. This constitutive equation requires the flux to be in the direction of the negative gradient; we say the contaminant "flows down the concentration gradient." It is the analog of Fourier's law in heat transfer, which states that the heat flux is proportional to the temperature gradient and heat flows from hotter to colder regions. As we shall observe in detail in Chapter 2, if there is bulk movement, or transport, of the liquid carrying the chemical solute, then there is a contribution to the flux that measures the dispersion of the contaminant due to the velocity variations of the fluid through the tortuous pathways in the porous medium. Here, one must take care not to interpret the term dispersion in the sense of dispersive waves, as in the theory of wave propagation. In porous media, the term has a physical connotation meaning "spreading out."

In three dimensions, the diffusion equation is

$$u_t = D\Delta u,$$

where $u = u(x, y, z, t)$ and Δ is the Laplacian operator in three dimensions given by

$$\Delta u = u_{xx} + u_{yy} + u_{zz}.$$

The three-dimensional diffusion equation arises from the conservation equation $u_t = -\nabla \cdot Q$ and Fick's law $Q = -D\nabla u$, where Q is the vector flux.[1] Recall that the gradient of u points in the direction of maximum increase, so again the flux is "down the gradient."

Equation (1.1) has some basic properties that are shared by other linear, parabolic partial differential equations.

(a) It propagates signals at infinite speed.

(b) It smooths out roughness in the initial or boundary data.

(c) There is loss of information as signals propagate.

(d) There is a maximum principle.

Much of the subsequent discussion in this chapter centers around these basic properties.

The simplest solutions of (1.1) are **plane wave solutions** of the form $u = \exp i(kx + \omega t)$, where k is the **wave number** and ω is the **frequency**. Substituting this form into (1.1) gives the dispersion relation $\omega = ik^2 D$, and therefore we obtain solutions of the form

$$u(x, t) = e^{-k^2 Dt}e^{ikx}.$$

---

[1]Generally we shall not represent vectors by special symbols or type; whether a quantity is a scalar or vector will be clear from context. The divergence of a vector field Q is denoted by $\nabla \cdot Q$ or div Q; the gradient of a scalar field u is denoted by ∇u or grad u.

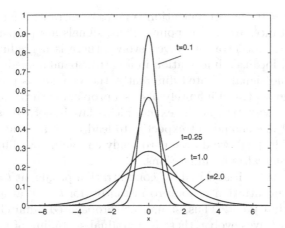

Figure 1.1: Time profiles of the fundamental solution showing the diffusion of a unit amount of contaminant released initially at $x = 0$.

Thus, initial, periodic wave profiles decay with decay rate $k^2 D$, which depends on the wave number k. Hence, higher-frequency waves decay more rapidly than lower-frequency waves. The reader might be reminded that the wave number represents the number of spatial oscillations per 2π length units ($2\pi/k$ is the wavelength), and the temporal period is $2\pi/\omega$.

The most important solution of (1.1) is the **fundamental solution** given by

$$u = g(x, t) = \frac{1}{\sqrt{4\pi Dt}} e^{-x^2/4Dt}. \tag{1.2}$$

This solution can be found by superimposing all of the plane wave solutions over all wave numbers, that is,

$$\int_{-\infty}^{\infty} e^{-k^2 Dt} e^{ikx} dk = \frac{1}{\sqrt{4\pi Dt}} e^{-x^2/4Dt}.$$

Note that the left-side of this expression is just the Fourier transform of the Gaussian function $e^{-k^2 Dt}$.

Exercise 1 *Derive the preceding integral formula.*

The fundamental solution (1.2) is the concentration surface that arises from an initial, unit contaminant source localized at the origin in a one-dimensional domain. The time cross sections, or time snapshots, of the concentration surface for different times, have the form of a bell-shaped curve. Interpreted in a probabilistic way, the time snapshots are normal probability densities of variance $2Dt$. They illustrate the properties listed above. See figure 1.1.

Because we think of diffusion in terms of the random motion and collisions of molecules, it is not surprising that the fundamental solution is a normal probability density. Even though the initial signal, or concentration profile, at $t = 0$

is localized at $x = 0$, the concentration is nonzero instantly (for $t > 0$) at arbitrarily large distances from that point. Thus, signals are propagated infinitely
fast. Second, as concentration waves evolve, there is a gradual spreading of
the wave profile; localized information in a narrow band spreads out until there
is no recognizable signal. Stated differently, the variance is proportional to t.
The fact that signals travel infinitely fast is a problem with the idealized model.
Although the conservation law is exact, Fick's law is not. It is an empirical,
local result, and one should not expect it to lead to an accurate description in
large systems. Although we do use it to study contaminant diffusion in infinite
systems, we must realize its limitations.

Even though the initial signal, or concentration profile, at $t = 0$ is localized
at $x = 0$, the concentration is nonzero instantly (for $t > 0$) at arbitrarily large
distances from that point. Thus, signals are propagated infinitely fast. Second,
as concentration waves evolve, there is a gradual spreading of the wave profile;
localized information in a narrow band spreads out until there is no recognizable signal. Stated differently, the variance is proportional to t. The fact that
signals travel infinitely fast is a problem with the idealized model. Although
the conservation law is exact, Fick's law is not. It is an empirical, local result,
and one should not expect it to lead to an accurate description in large systems.
Although we do use it to study contaminant diffusion in infinite systems, we
must realize its limitations.

As just noted, the fundamental solution $u = g(x, t)$ of (1.1) is the concen-
tration resulting from an initial, unit contaminant source localized at $x = 0$.
As such, g is the **Green's function** associated with the diffusion operator on
$\mathbb{R} = (-\infty, \infty)$. Indeed, it is easy to verify that

$$\int_{-\infty}^{\infty} g(x, t)dx = 1, \quad t > 0,$$

and $g(x, t)$ approaches the delta distribution $\delta(x)$ as $t \to 0^+$. If the contaminant
source is shifted to $x = \xi$ and has magnitude $\phi(\xi)$, rather than unity, then the
resulting concentration surface is $g(x - \xi, t)\phi(\xi)$. As such, if there is an initial
distributed concentration $\phi(\xi)$, at each $\xi \in \mathbb{R}$, then we may superimpose the
effects for each point source to obtain $u(x, t) = \int_{-\infty}^{\infty} g(x - \xi, t)\phi(\xi)d\xi$, which is
the concentration surface resulting from an initial, distributed concentration ϕ.
Thus, we have solved the **Cauchy problem**, or pure initial value problem, for
the dispersion equation:

$$
\begin{aligned}
u_t &= Du_{xx}, \quad x \in \mathbb{R}, \, t > 0 & (1.3) \\
u(x, 0) &= \phi(x), \quad x \in \mathbb{R}. & (1.4)
\end{aligned}
$$

We record the result formally.

Theorem 2 *Let ϕ be a bounded, continuous function defined on $\mathbb{R}$. Then the*
function

$$u(x, t) = \int_{-\infty}^{\infty} g(x - \xi, t)\phi(\xi)d\xi \tag{1.5}$$

is the unique, bounded solution to the Cauchy problem (1.3)–(1.4). The solution has continuous derivatives of all order in $t > 0$.

Exercise 3 *Using a simple change of variables, verify that the solution (1.5) to the Cauchy problem can be written*

$$u(x,t) = \frac{1}{\sqrt{\pi}} \int_{-\infty}^{\infty} e^{-r^2} \phi(x - r\sqrt{4Dt}) dr.$$

*This integral formula is called the **Poisson representation** of the solution to (1.3)–(1.4).*

If the initial datum ϕ is a bounded, *piecewise* continuous function, then (1.5) is the unique solution of (1.3) for $t > 0$ and the initial condition is satisfied in the sense

$$\lim_{t \to 0^+} u(x,t) = \frac{1}{2}(\phi(x+) + \phi(x-)).$$

That is, $u(x,t)$ approaches the average value of the left and right limits of the initial function as $t \to 0$. The solution still has continuous derivatives of all order in $t > 0$. Stated differently, the diffusion equation smooths out discontinuities, singularities, and variations in the initial profile. Two distinguished concentration profiles become indistinguishable after a long time.

Exercise 4 *If the initial datum is a step function given by $\phi(x) = 0$, $x < 0$, and $\phi(x) = 1$, $x > 0$, show that the solution to the Cauchy problem can be written*

$$u(x,t) = \frac{1}{2}\left(1 + \mathrm{erf}\left(\frac{x}{\sqrt{4Dt}}\right)\right),$$

where

$$\mathrm{erf}(z) = \frac{2}{\sqrt{\pi}} \int_0^z e^{-r^2} dr$$

*is the **error function**. Show that the spatial derivative of this solution is the fundamental solution, i.e., $u_x = g(x,t)$.*

Exercise 5 *Using the fact that the integral of the fundamental solution is unity, show*

$$|u(x,t)| \leq \max|\phi|, \quad x \in \mathbb{R}, \quad t > 0.$$

Hence, the solution is bounded by largest value of the initial data, which is the simplest version of a maximum principle.

We note that the solution (1.5) is the convolution (over x) of the fundamental solution and the initial condition. That is,

$$u(x,t) = g(x,t) * \phi(x) \equiv \int_{-\infty}^{\infty} g(x - \xi, t)\phi(\xi) d\xi.$$

Thus the solution to the Cauchy problem is the weighted average of the initial concentration profile with translates of the fundamental solution. In applied analysis, the convolution operation is usually regarded as a solution operator that maps the initial profile into the solution profile at time t.

Exercise 6 *In the Cauchy problem (1.3)–(1.4) let $\int \phi dx = 1$, $D = 1$, and $\phi \geq 0$. Show that the mean and variance of u, regarded as a probability density and which are given by*

$$\mu(t) = \int_R xu(x,t)dx, \qquad \sigma^2(t) = \int_R (x - \mu)^2 u(x,t)dx,$$

evolve according to

$$\mu(t) = \mu(0), \quad \sigma^2(t) = 2t + \sigma^2(0).$$

Thus, the mean remains constant, but the standard deviation increases with time. Therefore, the band width widens and information is lost.

Exercise 7 *When the initial datum is Gaussian, i.e., $\phi(x) = e^{-x^2}$, show that the solution to the Cauchy problem is*

$$u(x,t) = \frac{1}{\sqrt{1 + 4Dt}} e^{-x^2/(1+4Dt)}.$$

Exercise 8 *Another result indicating the loss of information deals with entropy changes. Consider the nonhomogeneous dispersion equation*

$$u_t = Du_{xx} + f(x,t), \quad 0 < x < l, \ t > 0$$

where f is a given source. Suppose u is a positive solution. If we define

$$S(t) \quad = \quad \int_0^l \ln u \, dx = \text{net entropy in } [0,l],$$

$$\Phi(t) \quad = \quad -D\frac{\partial}{\partial x} \ln u \, |_0^l = \text{net entropy flux through the boundary,}$$

$$\delta Q \quad = \quad \int_0^l \frac{f}{u} dx = \text{entropy increase due to source,}$$

prove that

$$\frac{dS}{dt} + \Phi \geq \delta Q,$$

which is one form of the Clasuius–Duhem inequality. [Hint: divide the equation by u and integrate, noting that $u_t/u = (\ln u)_t$ and $u_{xx}/u = (u_x/u)_x + (u_x/u)^2$].

Exercise 9 *If $v(x,t)$ satisfies the one-dimensional diffusion equation, show that $u(x,y,t)=v(x,t)v(y,t)$ satisfies the two-dimensional diffusion equation*

$$u_t = D(u_{xx} + u_{yy}).$$

Deduce that the fundamental solution to the diffusion equation in two dimensions is

$$g(r,t) = \frac{1}{4\pi t} e^{-r^2/4Dt}, \quad r^2 = x^2 + y^2.$$

In three dimensions the fundamental solution is

$$g(r,t) = \frac{1}{(4\pi t)^{3/2}} e^{-r^2/4Dt}, \quad r^2 = x^2 + y^2 + z^2.$$

A general second-order partial differential equation (PDE) in two independent variables of the form

$$a(x,t)u_{xx} + b(x,t)u_{xt} + c(x,t)u_{tt} = \psi(x,t,u,u_x,u_t) \qquad (1.6)$$

is **parabolic** at a point in spacetime (xt-space) if $b^2 - 4ac = 0$ at that point. It is
parabolic in a domain if it is parabolic at each point of the domain. Clearly the
diffusion equation (1.1) is parabolic in all of $\mathbb{R}^2$. As it turns out, every parabolic
equation (1.6) can be transformed, via a change of independent variables

$$\xi = \xi(x,t), \quad \eta = \eta(x,t)$$

into a PDE containing a single second derivative, i.e., an equation having the
form

$$u_{\eta\eta} = \Psi(\xi,\eta,u,u_\xi,u_\eta). \qquad (1.7)$$

This latter equation is the **canonical form** associated with (1.6).

Exercise 10 *To reduce (1.6) to the canonical form (1.7), if $a \neq 0$ show that
one can always choose $\xi = \xi(x,t)$ to be the integral curves $\xi(x,t) =$ constant of
the ordinary differential equation $dt/dx = b(x,t)/2a(x,t)$ and $\eta = \eta(x,t)$ to be
independent of ξ. A good choice is often $\eta = t$.*

Example 11 *Consider the parabolic equation*

$$x^2 u_{xx} + 2x u_{xt} + u_{tt} = u_t.$$

The integral curves of $dt/dx = b/2a = 1/x$ are $xe^{-t} = C$. Therefore choose

$$\xi = xe^{-t}, \quad \eta = t.$$

One can easily check that the equation becomes, in these new coordinates,

$$u_{\eta\eta} = u_\eta.$$

By direct integration this equation has the general solution

$$u = F(\xi) + G(\xi)e^{-\eta},$$

where F and G are arbitrary functions. Thus,

$$u(x,t) = F(xe^{-t}) + G(xe^{-t})e^{-t}$$

is the general solution to the given equation.

A parabolic equation that arises in many hydrogeological applications has
the form

$$u_t = (a(x,t)u_x)_x + F(x,t,u,u_x), \qquad (1.8)$$

where $a(x,t) > 0$. If $a(x,t)$ is bounded away from zero, i.e., $a(x,t) \geq a_0 > 0$,
then we say that (1.8) is **uniformly parabolic**. If $a(x,t)$ is not bounded away

from zero, we say (1.8) is **degenerate**. Degenerate equations, where the diffusion vanishes, lead to interesting solution properties not found for uniformly parabolic equations; for example, degenerate equations can propagate discontinuities. If $a(x,t)$ is ever negative, then the problem is ill-posed in the sense that arbitrarily small differences in two sets of initial data can evolve into large differences in the corresponding solutions; that is, stability, or continiuity with respect to initial data, is lost. In hydrogeology, $a(x,t)$, which represents a diffusion coefficient, is always nonnegative.

1.2   Semi-Infinite Domains

1.2.1   Boundary Value Problems

In many contaminant transport problems the porous domain is long and contains a single inlet boundary. Such problems are often modeled on a semi-infinite region $x > 0$ with the inlet boundary at $x = 0$.

The solution to the Cauchy problem obtained in the last section can be used to obtain solutions on semi-infinite domains by *reflection through a boundary*. Consider the **Cauchy–Dirichlet problem**, or the initial boundary value problem for the diffusion equation on a semi-infinite domain:

$$u_t = Du_{xx}, \quad x > 0, t > 0, \tag{1.9}$$

$$u(0,t) = 0, \quad t > 0, \tag{1.10}$$

$$u(x,0) = \phi(x), \quad x > 0. \tag{1.11}$$

A boundary condition of the form (1.10), where is concentration is specified at the boundary, is called a **Dirichlet boundary condition**. To solve this problem we use the method of reflection. We extend the problem (1.9)–(1.11) to the entire real axis by extending the initial data ϕ to an *odd function ψ* defined by

$$\psi(x) = \phi(x) \ \ if \ \ x > 0; \quad \psi(x) = -\phi(-x) \ \ if \ \ x < 0; \quad \psi(0) = 0.$$

We then solve the extended problem on $\mathbb{R}$ by formula (1.5) and then restrict that solution to the positive real axis, which is then the desired solution to the Cauchy–Dirichlet problem.

To this end let us consider the Cauchy problem

$$v_t = Dv_{xx}, \quad x \in \mathbb{R}, \quad t > 0,$$

$$v(x,0) = \psi(x), \quad x \in \mathbb{R},$$

where ψ is the odd extension of the function ϕ as defined above. By formula (1.5) the solution is given by

$$v(x,t) = \int_{-\infty}^{\infty} g(x - y, t)\psi(y)dy.$$

Breaking up this integral into two parts, $y < 0$ and $y > 0$, we obtain

$$
\begin{aligned}
v(x,t) &= \int_{-\infty}^{0} g(x-y,t)\psi(y)dy + \int_{0}^{\infty} g(x-y,t)\psi(y)dy \\
&= -\int_{-\infty}^{0} g(x-y,t)\phi(-y)dy + \int_{0}^{\infty} g(x-y,t)\phi(y)dy \\
&= \int_{0}^{\infty} [g(x-y,t) - g(x+y,t)]\phi(y)dy.
\end{aligned}
$$

We restrict this solution to $x > 0$, and therefore obtain the solution to the Cauchy–Dirichlet problem (1.9)–(1.11). Formally we record the result.

Theorem 12 *Let ϕ be a bounded, continuous function on $[0,\infty)$ with $\phi(0) = 0$. The solution to the Cauchy–Dirichlet problem (1.9)–(1.11) is given by*

$$
u(x,t) = \int_{0}^{\infty} [g(x-y,t) - g(x+y,t)]\phi(y)dy.
$$

In many practical problems we may not have the condition $\phi(0) = 0$, and ϕ may be only piecewise continuous. The theorem still holds in this case with the proviso that the solution approaches the average values of the left and right limits of ϕ as we approach the boundary.

The **Cauchy–Neumann problem**, where the Dirichlet boundary condition (1.10) at $x = 0$ is replaced by the **Neumann boundary condition** $u_x(0,t) = 0$, can be solved using the method of reflection by extending the initial data to an even function on the entire real line. This boundary condition is a zero-flux condition which does not permit solutes to cross the boundary.

Theorem 13 *The solution to the Cauchy–Neumann problem*

$$
\begin{aligned}
u_t &= Du_{xx}, \quad x > 0, t > 0 \\
u_x(0,t) &= 0, \quad t > 0, \\
u(x,0) &= \phi(x), \quad x > 0,
\end{aligned}
$$

is given by

$$
u(x,t) = \int_{0}^{\infty} [g(x-y,t) + g(x+y,t)]\phi(y)dy.
$$

Exercise 14 *Use the method of reflection to derive the solution to the Cauchy–Neumann problem given in Theorem 13.*

The solutions to the Cauchy–Dirichlet and Cauchy–Neumann problems can also be obtained using sine and cosine transforms.

Example 15 *(**Periodic Boundary Condition**) There is a natural problem on $x > 0$ where the input concentration at the boundary $x = 0$ is a periodic function of time, and that boundary condition has been applied for a long time.*

This situation may occur, for example, when there is seasonal runoff at an inlet boundary or the periodic dumping of contaminant wastes at the inlet boundary. The model for this problem is **Fourier–Kelvin problem**

$$u_t = Du_{xx}, \quad x > 0, \ t \in \mathbb{R},$$
$$u(0,t) = f(t), \quad t \in \mathbb{R},$$

where $f(t + P) = f(t)$ for all t, and $P > 0$ is the period. Because there is no initial condition, for well-posedness we must assume a condition at $x = \infty$ such as $u(x,t) \to 0$ as $x \to \infty$. For example, if $f(t) = \exp(i\omega t)$, where $P = 2\pi/\omega$, then we can assume a solution of the form $u = y(x)\exp(i\omega t)$. Thus, $i\omega y = Dy''$ with $y(0) = A$ and $y \to 0$ as $x \to \infty$. Then

$$y(x) = Ae^{-\sqrt{\omega/2D}(1+i)x},$$

which gives

$$u(x,t) = Ae^{-\sqrt{\omega/2D}x}e^{i\omega(t-\sqrt{1/(2D\omega)}x)}.$$

Note that the decay rate is proportional to the square root of ω and the concentration experiences a phase shift of amount $\sqrt{1/(2D\omega)}x$ at depth x.

Exercise 16 *For the Fourier–Kelvin problem take*

$$f(t) = \sum_{n=-\infty}^{\infty} a_n e^{in\omega t}, \quad a_n = \frac{1}{P}\int_0^P f(t)e^{-in\omega t}dt.$$

Show that the solution is given by

$$u(x,t) = \sum_{n=-\infty}^{\infty} a_n e^{in\omega t} \exp\left(-\sqrt{\frac{|n|\omega}{2D}}x - i\,\mathrm{sgn}(n)\sqrt{\frac{|n|\omega}{2D}}x\right).$$

Exercise 17 *Consider the initial boundary value problem*

$$u_t = Du_{xx}, \quad x > 0, t > 0, \tag{1.12}$$
$$u_x(0,t) - hu(0,t) = r(t), \quad t > 0, \tag{1.13}$$
$$u(x,0) = \phi(x), \quad x > 0, \tag{1.14}$$

where u is bounded. A boundary condition of the form (1.13) is called a **Fourier-type** *condition (also a Robin condition, or a radiation condition). Show that changing dependent variables to $v = u_x - hu$ transforms the Fourier condition to a Dirichlet condition, and verify that*

$$u(x,t) = e^{hx}\int_\infty^x e^{-h\xi}v(\xi,t)d\xi,$$

where v satifies the problem

$$v_t = Dv_{xx}, \quad x > 0, t > 0,$$
$$v(0,t) = r(t), \quad t > 0,$$
$$v(x,0) = \phi'(x) - h\phi(x), \quad x > 0.$$

Many problems on semi-infinite domains can also be solved by Laplace transform methods, a common method used in hydrogeology. Consider the diffusion of a contaminant into a clean medium $x > 0$ with diffusion constant D, where the contaminant is maintained at concentration $f(t)$ on the boundary. The model is

$$u_t = Du_{xx}, \quad x, t > 0, \tag{1.15}$$
$$u(x, 0) = 0, \quad x > 0, \tag{1.16}$$
$$u(0, t) = f(t), \quad t > 0. \tag{1.17}$$

If $U(x, s)$ denotes the Laplace transform of u on time, i.e.,

$$(\mathcal{L}u)(x, s) = U(x, s) = \int_0^\infty u(x, t)e^{-st}dt,$$

then the basic operational formulas dictating how derivatives transform are

$$(\mathcal{L}u_t)(x, s) = sU(x, s) - u(x, 0),$$
$$(\mathcal{L}u_{xx})(x, s) = U_{xx}(x, s).$$

That is, a time derivative is transformed into a multiplication operation (by s) in the transform domain, and spatial derivatives are unaffected. Taking the Laplace transform of the PDE therefore gives

$$sU = DU_{xx},$$

which has bounded solution

$$U(x, s) = A(s)e^{-\sqrt{s/D}x},$$

where A is an arbitrary function. Taking the Laplace transform of the boundary condition gives $U(0, s) = F(s)$, where F is the transform of f. Thus, $A(s) = F(s)$, and so the solution to the problems in the transform domain is

$$U(x, s) = F(s)e^{-\sqrt{s/D}x}.$$

The convolution theorem allows us to invert the product of Laplace transforms on the right-side. The convolution theorem states

$$\mathcal{L}^{-1}(F(s)H(s)) = (f * h)(t),$$

where

$$(f * h)(t) = \int_0^t f(\tau)h(t - \tau)d\tau$$

is the convolution of f and h. Equivalently, the Laplace transform of a convolution is the product of the Laplace transforms, i.e., $(\mathcal{L}(f * h))(s) = F(s)G(s)$

[see, for example, Schiff (1999)]. The solution to (1.15)–(1.17) is therefore

$$
\begin{aligned}
u(x,t) &= f(t) * \mathcal{L}^{-1}\left(e^{-\sqrt{s/D}x}\right) \\
&= f(t) * \frac{x}{t}g(x,t) \\
&= \int_0^t \frac{x}{t-\tau}g(x,t-\tau)f(\tau)d\tau \\
&= \int_0^t \frac{x}{\sqrt{4\pi D(t-\tau)^3}}e^{-x^2/4D(t-\tau)}f(\tau)d\tau,
\end{aligned}
$$

where g is the fundamental solution. In the second equality we used a table of Laplace transforms to find $\mathcal{L}^{-1}\left(e^{-\sqrt{s/D}x}\right) = \frac{x}{t}g(x,t)$.

1.2.2   An Inverse Problem

As a final exercise, let us set up and solve a parameter identification problem, or an **inverse problem**. Consider a long ($x \geq 0$) porous domain characterized by an unknown diffusion constant D. Suppose that measurements can only be made at the inlet boundary $x = 0$. If a known chemical tracer concentration $f(t)$ at $x = 0$ is applied, can we measure the tracer flux at $x = 0$ at a single instant of time t_0 and thereby determine D? We assume $f(0) = 0$. The PDE model for this problem is

$$
\begin{aligned}
u_t &= Du_{xx}, \quad x,t > 0, & (1.18) \\
u(0,t) &= f(t), \quad t > 0, & (1.19) \\
u(x,0) &= 0, \quad x > 0, & (1.20)
\end{aligned}
$$

which is a well-posed problem. We are asking if we can determine D from a single flux measurement

$$
-Du_x(0,t_0) = a, \tag{1.21}
$$

where a is known. According to the result in the last section, the solution to the problem (1.18)–(1.20) is

$$
u(x,t) = \int_0^t \frac{x}{\sqrt{4D\pi(t-\tau)^3}}e^{-x^2/4D(t-\tau)}f(\tau)d\tau. \tag{1.22}
$$

It appears that the strategy should be to calculate the flux at $(0,t_0)$ from the solution formula. Indeed this is the case, but calculating the x derivative of u is not a simple matter. The straightforward approach of pulling a partial derivative $\partial/\partial x$ under the integral sign fails because one of the resulting improper integrals cannot be evaluated at $x = 0$; it does not exist and therefore we must be more clever in calculating u_x. To this end we note that the solution (1.22) can be written as

$$
u(x,t) = -2D\int_0^t \frac{\partial}{\partial x}g(x,t-\tau)f(\tau)d\tau,
$$

where $g(x,t)$ is the fundamental solution

$$g(x,t) = \frac{1}{\sqrt{4Dt}}e^{-x^2/4Dt}.$$

Therefore,

$$
\begin{aligned}
u_x(x,t) &= -2D \int_0^t \frac{\partial^2}{\partial x^2} g(x, t-\tau) f(\tau) d\tau \\
&= 2 \int_0^t \frac{\partial}{\partial \tau} g(x, t-\tau) f(\tau) d\tau,
\end{aligned}
$$

because $-g_\tau = g_t = Dg_{xx}$. Now we integrate by parts to obtain

$$u_x(x,t) = -2 \int_0^t g(x, t-\tau) f'(\tau) d\tau.$$

The boundary terms generated by the integration by parts are zero. Consequently, we have, by (1.21),

$$-Du_x(0, t_0) = \sqrt{D} \int_0^{t_0} \frac{f'(\tau)}{\sqrt{\pi(t_0 - \tau)}} d\tau = a.$$

This equation uniquely determines the diffusion constant D and solves the parameter identification problem. For example, if $f(t) = \beta t$, i.e., the concentration is increased linearly, then the integral can be calculated exactly and we obtain $D = \frac{\pi a^2}{4\beta^2 t_0}$.

1.3 Sources and Duhamel's Principle

Sources are common in hydrogeological phenomena. For example, both precipitation and irrigation can carry contaminants to the subsurface, distributing solutes over the entire domain of interest. As we noted already, the fundamental solution gives the solution when the source is a point source given at time t=0. Now we look at the case when the source is distributed in both space and time.

The Cauchy problem for the diffusion equation on $\mathbb{R}$ with a source term that depends on both location and time is

$$
\begin{aligned}
u_t &= Du_{xx} + f(x,t), \quad x \in \mathbb{R}, \quad t > 0, & (1.23) \\
u(x, 0) &= 0, \quad x \in \mathbb{R}. & (1.24)
\end{aligned}
$$

Here f is a given contaminant source. The key to obtaining the solution to problems with sources is **Duhamel's principle,** which states, in general terms, that the solution to an evolution problem with a source can be obtained from the solution of the initial value problem without a source. In other words, the homogeneous problem contains the seed for the solution to the nonhomogeneous problem.

For ordinary differential equations Duhamel's principle may be stated simply as follows: The solution $y = y(t)$ of the ODE model

$$y'(t) = ay + f(t), \quad t > 0; \quad y(0) = 0 \tag{1.25}$$

is given by

$$y(t) = \int_0^t w(t - \tau, \tau)d\tau,$$

where $w = w(t, \tau)$ is the solution of the homogeneous problem

$$w'(t, \tau) = aw(t, \tau), \quad t > 0; \quad w(0, \tau) = f(\tau).$$

Here we are introducing τ as a new parameter. This is clearly correct because the last equation has solution $w(t, \tau) = f(\tau)e^{at}$, and therefore equation (1.25) has solution

$$y(t) = \int_0^t e^{a(t-\tau)} f(\tau)d\tau.$$

We apply this same principle to the diffusion problem (1.23)–(1.24). In this case, the solution is

$$u(x, t) = \int_0^t w(x, t - \tau, \tau)d\tau,$$

where $w(x, t; \tau)$ solves the homogeneous problem

$$w_t = Dw_{xx}, \quad x \in \mathbb{R}, \quad t > 0, \tag{1.26}$$
$$w(x, 0; \tau) = f(x, \tau), \quad x \in \mathbb{R}. \tag{1.27}$$

One can easily verify this fact. Then, we can write down the explicit formula. By (1.5) the solution to (1.26)–(1.27) is

$$w(x, t; \tau) = \int_{-\infty}^{\infty} g(x - y, t)f(y, \tau)dy,$$

where g is the fundamental solution. Therefore, we have the following result.

Theorem 18 *The solution to the nonhomogeneous problem (1.23)–(1.24) is given by*

$$u(x, t) = \int_0^t \int_{-\infty}^{\infty} g(x - y, t - \tau)f(y, \tau)dyd\tau.$$

Using superposition, we may now write down the formula for the solution to the problem

$$u_t = Du_{xx} + f(x, t), \quad x \in \mathbb{R}, t > 0 \tag{1.28}$$
$$u(x, 0) = \phi(x), \quad x \in \mathbb{R}, \tag{1.29}$$

which has both a contaminant source and an initial contaminant concentration. By linearity, the solution to (1.28)–(1.29) is the sum

$$u(x,t) = \int_{-\infty}^{\infty} g(x-y,t)\phi(y)dy + \int_0^t \int_{-\infty}^{\infty} g(x-y,t-\tau)f(y,\tau)dyd\tau. \quad (1.30)$$

This is the **variation of parameters formula** for the problem (1.28)–(1.29).

We end this section with some general remarks on the viewpoint that applied mathematical analysts take regarding initial value problems with source terms. Applied scientists may be unfamiliar with this more abstract way of looking at such problems, but it is important because investigations couched in this formalism do appear in applied mathematics literature. The viewpoint is mostly a change of language to that of the qualitative theory of PDEs. To begin, in terms of convolutions, the variation of parameters formula (1.30) is

$$u(x,t) = g(x,t) \star \phi(x) + \int_0^t g(x,t-\tau) \star f(x,\tau)d\tau.$$

If we introduce a solution operator $S(t)$ defined by the action

$$S(t)\phi \equiv g(\cdot,t) \star \phi,$$

then the solution formula can be written

$$u(x,t) = S(t)\phi(x) + \int_0^t S(t-\tau)f(x,\tau)d\tau.$$

The first term of this representation is called the **free solution**, and the second term is the **nonhomogeneous solution**. We can think of the operator $S(t)$ as mapping the initial concentration profile $\phi(x)$ to the solution profile at time t. This formula clearly shows that the nonhomogeneous solution can be constructed from the free solution. To go one step further, if we supress the spatial variable x and write $u = u(t)$ as a concentration *profile* at time t (because it is a profile, it is automatically a function of x), then the variation of parameters formula is simply

$$u(t) = S(t)\phi + \int_0^t S(t-\tau)f(\cdot,\tau)d\tau.$$

This equation is reminiscent of the variation of parameters formula

$$y(t) = e^{at}y_0 + \int_0^t e^{a(t-\tau)}f(\tau)d\tau$$

for the solution of the initial value problem

$$y' = ay + f(t), \quad t > 0; y(0) = y_0$$

for an ordinary differential equation with $S(t)$ identified as e^{at}. Indeed, if we surpress x and write the initial value problem (1.28)–(1.29) as

$$u_t = Au + f(\cdot, t), \quad t > 0, \tag{1.31}$$
$$u(0) = \phi, \tag{1.32}$$

where A is the second-order differential operator $A \equiv D\, d^2/dx^2$, then it is tempting to identify an exponential operator e^{At} with the solution operator $S(t)$. Then we can write the solution to (1.28)–(1.29), or equivalently (1.31)–(1.32), as

$$u(t) = e^{At}\phi + \int_0^t e^{A(t-\tau)} f(\tau) d\tau.$$

Observe that this formula is precisely the formula that we would write down for the solution to a linear system of ODEs

$$u'(t) = Au(t) + f(t), \quad u(0) = \phi,$$

where $u(t) = (u(t), ..., u(t))$ is the unknown vector function, A is an n by n matrix, ϕ is the initial state vector, and $f(t)$ is a given vector; then e^{At} is a fundamental solution (recall that e^{At} is defined by $e^{At} = I + At + \frac{1}{2!}A^2 t^2 + \cdots$ when A is a matrix). Consequently, in this general language, we get the same formula regardless of whether we have a scalar ODE, a system of ODEs, or a PDE. A partial differential equation can therefore be viewed as an ordinary differential equation in an infinite-dimensional function space (the space of possible contaminant profiles), and many of the concepts formulated for ODEs (in finite-dimensional spaces) carry over in a natural way to PDEs, including the definition of an exponential e^{At} when A is a differential operator. Thus, equation (1.31) is regarded as an ODE that governs the evolution of the wave profile $u = u(t)$. Viewed in this manner, PDEs have just a cosmetic difference from ODEs—from finite-dimensional objects to infinite-dimensional objects. ODEs evolve solutions in finite-dimensional spaces ($\mathbb{R}^n$), while PDEs evolve solutions in infinite-dimensional spaces.

1.4   Bounded Domains

It is generally true that solution methods for problems on infinite or semi-infinite spatial domains usually involve some kind of transform (Fourier, Laplace, Hankel, and so on), or some kind of reflection method.

The standard method for solving parabolic problems on bounded spatial domains is the method of **eigenfunction expansions** (also called Fourier's method or separation of variables). The idea is to seek the solution as a linear combination of eigenfunctions of a related Sturm–Liouville problem. This method should be familiar to the reader, and it is the method that is usually emphasized in elementary contexts. In this section, for completeness, we briefly review this method, but we shall give more emphasis to its mathematical structure. We denote the set of **square integrable functions** on an interval $[0, l]$ by $L^2[0, l]$. This is the set

$$L^2[0, l] = \{f \mid \int_0^l |f(x)|^2 dx < \infty\}.$$

The integral in this definition is the Lebesgue integral, but readers unfamiliar with the Lebesgue integral may regard it as the usual Riemann integral introduced in elementary calculus; all of the functions we encounter are Riemann integrable and therefore Lebesgue integrable. In $L^2[0, l]$ we introduce the **inner product** of f and g defined by

$$(f, g) = \int_0^l f(x)g(x)dx,$$

and we define

$$\|f\| = \sqrt{(f, f)} = \left(\int_0^l f(x)^2 dx \right)^{1/2}$$

as the **norm** of f that is associated with the inner product. This is sometimes called the energy norm. The inner product and norm put geometry on the infinite-dimensional function space $L^2[0, l]$ in much the same way that the dot product and norm (length) put geometry on the finite-dimensional vector space $\mathbb{R}^3$. With this inner product and associated norm, the space of functions $L^2[0, l]$ is a **Hilbert space**; that is, it is a linear space on which there is defined an inner product, and the space is **complete** in the norm generated by the inner product. This latter statement means that if a sequence of functions $f_n(x)$ in $L^2[0, l]$ has the property that $\|f_{n+p} - f_n\| \to 0$ as $n \to \infty$ for any $p \geq 0$, then it must converge; that is, there is an $f \in L^2[0, l]$ for which $\|f_n - f\| \to 0$ as $n \to \infty$. This is called the completeness property of $L^2[0, l]$.

The problem we consider is the initial boundary value problem

$$
\begin{aligned}
u_t &= Au + f(x, t), \quad x \in (0, l), \quad t > 0, & (1.33) \\
B_1 u(0, t) &\equiv \alpha_1 u(0, t) + \alpha_2 u_x(0, t) = 0, \quad t > 0, & (1.34) \\
B_2 u(l, t) &\equiv \beta_1 u(l, t) + \beta_2 u_x(l, t) = 0, \quad t > 0, & (1.35) \\
u(x, 0) &= \phi(x), \quad x \in [0, l], & (1.36)
\end{aligned}
$$

where A is the second-order, spatial, differential operator having the form

$$A \equiv -\frac{d}{dx}\left(p(x)\frac{d}{dx}\right) + q(x)\frac{d}{dx}.$$

We assume that p, q, and p' are continuous on $[0, l]$ and $p(x) \neq 0$ in $[0, l]$. If $\alpha_2 = 0$, then we say the boundary condition at $x = 0$ is **Dirichlet type**, and if $\alpha_1 = 0$ we say the boundary condition at $x = 0$ is **Neumann type**. If, say, both α_1 and α_2 are nonzero, then we say the boundary condition at $x = 0$ is a **Fourier** (or adsorption, or radiation) condition. We make similar statements at $x = l$. We always assume that a boundary condition does not degenerate.

Associated with the boundary value problem (1.33)–(1.36) is a natural, pure boundary value problem for $y = y(x)$ called a **regular Sturm–Liouville problem**, or **SLP**, defined by

$$
\begin{align}
Ay &= \lambda y, \quad x \in (0, l), \tag{1.37} \\
B_1 y(0) &= 0, \tag{1.38} \\
B_2 y(l) &= 0. \tag{1.39}
\end{align}
$$

We say that a constant λ is an **eigenvalue** for the differential operator A with given boundary conditions if there exists a nontrivial solution $y(x)$, called an **eigenfunction**, corresponding to that value of λ. The pair (λ, y) is called an **eigenpair**. The following theorem is basic.

Theorem 19 *The regular Sturm–Liouville problem (1.37)–(1.38) has infinitely many eigenvalues λ_n; the eigenvalues are real, $\lim |\lambda_n| = +\infty$, and the corresponding eigenfunctions $y_n(x)$ form an orthonormal set, i.e., $(y_m, y_n) = 0$ if $m \neq n$, and $\|y_n\| = 1$ for all n. Moreover, any $f \in L^2[0, l]$ can be expanded uniquely in a generalized Fourier series*

$$
f(x) = \sum_{n=1}^{\infty} (f, y_n) y_n(x),
$$

where convergence is in $L^2[0, l]$, i.e.,

$$
\lim_{N \to \infty} \int_0^l |f(x) - \sum_{n=1}^{N} (f, y_n) y_n(x)|^2 dx = 0.
$$

This last statement in the theorem means that the eigenfunctions form a basis for the space of square integrable functions. The coefficients (f, y_n) in the generalized Fourier series for f are called the Fourier coefficients.

To solve the boundary value problem (1.33)–(1.36), where ϕ and f are square integrable, we make the Ansatz

$$
u(x, t) = \sum c_n(t) y_n(x),
$$

where the y_n are the orthonormal eigenfunctions of the associated Sturm-Liouville problem (1.37)–(1.39), and the $c_n(t)$ are to be determined. All sums range over $n = 1$ to $n = \infty$. Also we expand the source function in terms of the eigenfunctions as

$$
f(x, t) = \sum f_n(t) y_n(x), \qquad f_n(t) = (f(\cdot, t), y_n).
$$

Substituting into the PDE (1.33) gives

$$
\begin{align}
\sum c_n'(t) y_n(x) &= \sum c_n(t) A y_n(x) + \sum f_n(t) y_n(x) \\
&= \sum c_n(t) \lambda_n y_n(x) + \sum f_n(t) y_n(x).
\end{align}
$$

Equating the coefficients of the independent eigenfunctions gives a system of uncoupled ODEs for the coefficients $c_n(t)$:

$$c_n' = \lambda_n c_n + f_n(t), \quad n = 1, 2, \ldots.$$

To obtain initial conditions we set

$$u(x, 0) = \sum c_n(0) y_n(x) = \phi(x) = \sum (\phi, y_n) y_n(x).$$

Whence, the $c_n(0)$ are the Fourier coefficients of the initial function, or

$$c_n(0) = (\phi, y_n), \quad n = 1, 2, \ldots.$$

By the variation of parameters formula, the solution to the initial value problem for c_n is

$$c_n(t) = (\phi, y_n) e^{\lambda_n t} + \int_0^t e^{\lambda_n(t-\tau)} f_n(\tau) d\tau.$$

Therefore, the solution to the boundary value problem is

$$u(x, t) = \sum (\phi, y_n) e^{\lambda_n t} y_n(x) + \sum \left(\int_0^t e^{\lambda_n(t-\tau)} f_n(\tau) d\tau \right) y_n(x). \qquad (1.40)$$

This calculation was completely formal, meaning it was done without rigorous verification of each step; the actual validity of the formula depends on the regularity of the functions ϕ and f. One can prove, for example, the following: If ϕ is a bounded, integrable function on $[0, l]$ and if f is twice continuously differentiable with respect to x for $x \in [0, l]$ with $B_1 f(0, t) = 0$, $B_2 f(l, t) = 0$, then the series (1.40) defines a solution [a function $u(x, t)$ for which u_t and u_{xx} are continuous] to (1.33) in $(0, l) \times (0, T)$ for any $T > 0$. Furthermore, if ϕ is twice continuously differentiable and $B_1 \phi(0) = 0$ and $B_2 \phi(l) = 0$, then (1.40) satisfies the initial and boundary conditions (1.34)–(1.36).

Many examples of this procedure can be found in the elementary texts listed in the references at the end of this book.

Exercise 20 *Show that the solution formula (1.40) can be written in the form*

$$u(x, t) = \int_0^l g(x, \xi, t) \phi(\xi) d\xi + \int_0^t \int_0^l g(x, \xi, t - \tau) f(\xi, \tau) d\xi d\tau,$$

where $g(x, \xi, t) = \sum e^{\lambda_n t} y_n(x) y_n(\xi)$. Show that this formula can also be obtained from Duhamel's principle.

1.5 The Maximum Principle

1.5.1 Maximum and Minimum Principles

A maximum (mininum) principle in differential equations is a statement about where the solution to a given equation takes on its maximum (minimum) value.

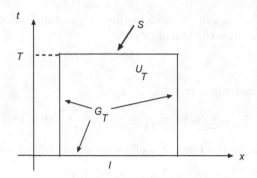

Figure 1.2: Spacetime domain for the maximum principle.

Diffusion processes, by their very physical nature, tend to smear out the concentration function, and this precludes clumping of the solution (the contaminant concentration) at interior points of the spacetime domain; recall, by Fick's law, the flux is proportional to the negative concentration gradient and there is flow *down the concentration gradient* or from higher to lower concentrations. Thus, the diffusion equation

$$u_t = Du_{xx}$$

has solutions whose maximum occurs initially or on the boundary of the spatial domain.

To be precise in the formulation, let $U_T = I \times (0, T)$ denote an open spacetime region, where I is an open, bounded spatial interval on the real line. Let G_T denote the part of the boundary of U_T consisting of the bottom and sides, and excluding the open segment $S = I \times \{T\}$ at $t = T$ (see figure 1.2).

We first formulate and prove the maximum principle for general parabolic operators of the form

$$Lu = a(x, t)u_{xx} + b(x, t)u_x - u_t,$$

where a and b are continuous on $\overline{U_T}$ and $a(x, t) \geq a_0 > 0$ on $\overline{U_T}$. We always assume $u \in C^2(U_T) \cap C(\overline{U_T})$. Here, we can regard $a(x, t)$ as a diffusion coefficient and $b(x, t)$ as a convection speed; we will meet the latter in Chapter 2.

Theorem 21 (Weak Maximum Principle). *If $Lu \geq 0$ in U_T, then the maximum value of u is attained on G_T. Similarly, if $Lu \leq 0$ in U_T, then the minimum value of u is attained on G_T.*

Proof. In the case of strict inequality, i.e., $Lu > 0$, the maximum cannot occur in U_T. If it did occur at a point P in U_T, then we would have $u_t = u_x = 0$ at P and $a(x, t)u_{xx} \leq 0$ at P. This contradicts $Lu > 0$. So the maximum of u must occur on $G_T \cup S$.

Now, if $Lu \geq 0$, then the same can be said, i.e., the maximum must occur on $G_T \cup S$. For, take $u_\epsilon = u + \epsilon e^{rx}$ where $\epsilon > 0$ is arbitrary. Then $Lu_\epsilon > 0$ and thus the maximum of u_ϵ must occur on $G_T \cup S$ by what we just proved. Then

$$\max_{\overline{U_T}} u \leq \max_{\overline{U_T}} u_\epsilon = \max_{G_T \cup S} u_\epsilon = \max_{G_T \cup S} u + C\epsilon.$$

Because ϵ is arbitrary, we have

$$\max_{\overline{U_T}} u \leq \max_{G_T \cup S} u.$$

Now, in fact, the maximum cannot occur on S. This is certainly true if we have strict inequality, i.e., $Lu > 0$. Otherwise, at a point P on S we would have $a(x,t)u_{xx} + b(x,t)u_x \leq 0$, which implies $u_t < 0$ at P, a contradiction. Now assume $Lu \geq 0$ and let $u_\epsilon = u + \epsilon e^{-t}$. Thus $Lu_\epsilon > 0$. By our remarks we know u_ϵ cannot have its maximum on S, so its maximum must occur on G_T. Then, as above,

$$\max_{\overline{U_T}} u \leq \max_{\overline{U_T}} u_\epsilon = \max_{G_T} u_\epsilon = \max_{G_T} u + \epsilon,$$

which yields

$$\max_{\overline{U_T}} u \leq \max_{G_T} u$$

This completes the proof. ∎

Note that this version of the maximum principle involves differential inequalities. In hydrogeology we are of course interested in the case of equality when we have a PDE.

Also, this weak version of the maximum principle asserts that the maximum of u over $\overline{U_T}$ must occur on G_T; this does not preclude the maximum also occurring in U_T. A stronger version of the maximum principle states that, in fact, if the maximum concentration occurs at (x_0, t_0) in U_T, then u must be a constant in U_{t_0}, i.e., a constant up to time t_0. A precise statement and proof are given later in this section.

Actually, we can include a linear decay source term $c(x,t)u$ in the operator L and obtain the same result. Let

$$Pu = a(x,t)u_{xx} + b(x,t)u_x + c(x,t)u - u_t,$$

where c is also continuous on the closure of U_T. Then we have:

Corollary 22 *If $Pu \geq 0$ in U_T, then the maximum value of u is attained on G_T. Similarly, if $Pu \leq 0$ in U_T, then the minimum value of u is attained on G_T.*

Proof. Note that $Mu = Lu + cu$. Let $u = ve^{rt}$. Then $e^{rt}(Lv - rv + cv) \geq 0$, or

$$Lv \geq (r - c)v$$

If we pick r large enough we can guarantee $r - c > 0$. Thus, $Lv \geq 0$, and v has it maximum on G_T. Thus, so does u, which is a positive multiple of v. ∎

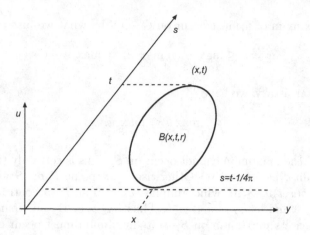

Figure 1.3: The diffusion ball $B(x, t, r)$.

The proof we present of the strong maximum principle for the diffusion equation follows the approach given in Evans (1998). What is important here is not the proof, but the idea behind the proof and the insights the idea brings to our understanding of diffusion. It employs a diffusion analog of the mean value property for harmonic functions. By way of motivation, the reader may recall that if $u = u(x, y)$ is a solution to Laplace's equation in an open domain Ω of the xy plane, then for any open disk $B_r(x, y)$ centered at (x, y) with radius r, which is contained in Ω,

$$u(x, y) = \frac{1}{\pi r^2} \int_{B_r(x,y)} u \, dA.$$

In other words, the value of u at the center of a disk is the average value of u over the entire disk. There is a similar *weighted* mean value property for solutions of the diffusion equation with an appropriately chosen "ball" whose boundary consists of level curves of the fundamental solution (taking $D = 1$)

$$g(x, t) = \frac{1}{\sqrt{4\pi t}} e^{-x^2/4t}.$$

We define the **diffusion ball** $B(x, t, r)$ at the point (x, t) of size $r > 0$ to be the set

$$B(x, t, r) = \{(y, s) \mid s < t, g(x - y, t - s) \geq \frac{1}{r}\}.$$

A diffusion ball is shown in figure 1.3.

Observe that (x, t) is at the top of the ball.

Then we have the following mean value property for the diffusion equation, from which a strong maximum principle will follow easily.

Theorem 23 *(Mean Value Property) Let $u = u(x,t)$ be a solution to the diffusion equation $u_t = u_{xx}$ in the spacetime domain U_T. Then*

$$u(x,t) = \frac{1}{4r} \int_{B(x,t,r)} u(y,s) \frac{(x-y)^2}{(t-s)^2} dy ds$$

for every $B(x,t,r)$ contained in U_T.

Proof. The outline of the proof is as follows. Because the diffusion equation is invariant under translations, we may as well translate the point (x,t) to the origin. Thus, we need to show

$$u(0,0) = \frac{1}{4r} \int_{B(0,0,r)} u(y,s) \frac{y^2}{s^2} dy ds.$$

To this end, let us use the notation $B(0,0,r) \equiv B(r)$ and define the function

$$\phi(r) = \frac{1}{r} \int_{B(r)} u(y,s) \frac{y^2}{s^2} dy ds = \int_{B(1)} u(ry, r^2 s) \frac{y^2}{s^2} dy ds.$$

Then we show $\phi'(r) = 0$, or $\phi(r) = \text{const.}$ Thus, $\phi(r)$ is a conserved quantity. In fact, we will show $\phi(r) = 4u(0,0)$, and we will be done.

Carrying out the details, we apply the chain rule to get

$$
\begin{aligned}
\phi'(r) &= \int_{B(1)} \left(u_y \frac{y^3}{s^2} + 2r u_s \frac{y^2}{s} \right) dy ds \\
&= \frac{1}{r^2} \int_{B(r)} \left(u_y \frac{y^3}{s^2} + 2 u_s \frac{y^2}{s} \right) dy ds \\
&\equiv J_1 + J_2.
\end{aligned}
$$

Now let

$$\psi \equiv -\frac{1}{2} \ln(-4\pi s) + \frac{y^2}{4s} + \ln r,$$

and note that $\psi \equiv 0$ on $\partial B(r)$. Using $\psi_y = y/2s$ we can write J_2 as

$$
\begin{aligned}
J_2 &= \frac{1}{r^2} \int_{B(r)} 4 u_s y \psi_y \, dy ds \\
&= -\frac{1}{r^2} \int_{B(r)} (4 u_s \psi + 4 u_{sy} y \psi) \, dy ds,
\end{aligned}
$$

where in the last step we used the identity $(4y u_s \psi)_y = 4 u_s \psi + 4 y \psi u_{sy} + 4 y u_s \psi_y$. Next we integrate the last integral by parts on s to obtain

$$
\begin{aligned}
J_2 &= \frac{1}{r^2} \int_{B(r)} (-4 u_s \psi + 4 u_y y \psi_s) \, dy ds \\
&= \frac{1}{r^2} \int_{B(r)} \left(-4 u_s \psi + 4 u_y y \left(-\frac{1}{2s} - \frac{y^2}{4s^2} \right) \right) dy ds \\
&= \frac{1}{r^2} \int_{B(r)} \left(-4 u_s \psi - \frac{2}{s} u_y y \right) dy ds - J_1.
\end{aligned}
$$

Because $u_s = u_{yy}$, we have

$$
\begin{aligned}
\phi'(r) &= \frac{1}{r^2} \int_{B(r)} \left(-4u_{yy}\psi - \frac{2}{s}u_y y \right) dy ds \\
&= \frac{1}{r^2} \int_{B(r)} \left(4u_y \psi_y - \frac{2}{s}u_y y \right) dy ds \\
&= 0.
\end{aligned}
$$

Consequently, ϕ is constant and

$$
\phi(r) = \phi(0) = u(0,0) \int_{B(1)} \frac{y^2}{s^2} dy ds = 4u(0,0).
$$

To see the last step we observe that the boundary of $B(1)$ is given by

$$
|y| = 2\sqrt{s \ln \sqrt{-4\pi s}}, \quad -\frac{1}{4\pi} < s < 0.
$$

Then

$$
\begin{aligned}
\int_{B(1)} \frac{y^2}{s^2} dy ds &= 2 \int_{-1/4\pi}^0 \left(\int_0^{2\sqrt{s \ln \sqrt{-4\pi s}}} y^2 dy \right) \frac{1}{s^2} ds \\
&= \frac{16}{3} \int_0^{1/4\pi} \frac{\sqrt{-\ln \sqrt{4\pi w}}^3}{\sqrt{w}} dw \\
&= \frac{16}{3\sqrt{\pi}} \int_0^1 \sqrt{-\ln v}^3 \, dv \\
&= \frac{16}{3\sqrt{\pi}} \int_0^1 \left(\ln \frac{1}{v} \right)^{3/2} dv \\
&= \frac{16}{3\sqrt{\pi}} \Gamma \left(\frac{5}{2} \right) \\
&= 4.
\end{aligned}
$$

This completes the argument. ∎

We can now us the mean value property to give a simple proof of the strong maximum principle.

Theorem 24 (Strong Maximum Principle) *Let u be continuous in the $\overline{U_T}$ and twice continuously differentiable in U_T, and suppose $u_t = u_{xx}$. Then*

$$
\max_{\overline{U_T}} u = \max_{G_T} u.
$$

Moreover, if there is a point $(x_0, t_0) \in U_T$ for which $u(x_0, t_0) = \max_{\overline{U_T}} u \equiv M$, then

$$
u = M \quad \text{in } U_{t_0}.
$$

Proof. By choosing r small enough, we can pick a diffusion ball $B(x_0, t_0, r)$ that lies entirely in U_T. Then the mean value property implies

$$M = u(x_0, t_0) = \frac{1}{r} \int_{B(x_0, t_0, r)} u(y, s) \frac{(y_0 - y)^2}{(t_0 - s)^2} dy ds \leq M.$$

The last step follows from the fact

$$\frac{1}{r} \int_{B(x_0, t_0, r)} \frac{(y_0 - y)^2}{(t_0 - s)^2} dy ds = 1.$$

This calculation is similar to the one given at the end of the proof of the mean value property. Thus, it follows that $u(y, s) = M$ for all (y, s) in the diffusion ball $B(x_0, t_0, r)$. Now pick any other point P in U_{t_0} and connect (x_0, t_0) to P with a straight line L. At the point (z_0, s_0) of intersection of L with the boundary of $B(x_0, t_0, r)$ we can put a diffusion ball $B(z_0, s_0, r)$ and conclude that $u = M$ in $B(z_0, s_0, r)$. Continuing in this manner we get $u = M$ on all of L, and in particular, at P. Thus, $u = M$ in U_{t_0}, and this completes the proof. ∎

The idea of considering a diffusion ball can be based upon a heuristic notion of the *effective* region of influence for the diffusion equation. If a unit point source is prescribed at $x = 0$ as initial data, we know the concentration for $t > 0$ is given by the fundamental solution

$$g(x, t) = \frac{1}{\sqrt{4\pi t}} e^{-x^2/4t}.$$

If we presuppose there is a sufficiently small concentration ϵ, below which concentrations are imperceptible, then the measurable concentrations are located in the set (x, t) for which $g(x, t) \geq \epsilon$. This region is the diffusion ball, as shown in the figure 1.4, and it is the effective region of influence for a point source at $x = t = 0$. Interpreted in this context, a signal governed by the diffusion equation does not propagate at infinite speed as we noted in Section 1, but at a "practical" finite speed.

We have not given the most general statement of the maximum principles. We refer the reader to the standard reference, Protter and Weinberger (1967), for a complete discussion of maximum principles in differential equations.

1.5.2 Comparison and Uniqueness

A comparison principle, a statement that compares solutions to similar problems, is a simple corollary of the maximum principle.

Theorem 25 *(Comparison Principle) Under the same hypotheses as the Weak Maximum Principle, if $Pu \geq Pv$ in U_T and $u \leq v$ on G_T, then $u \leq v$ on U_T.*

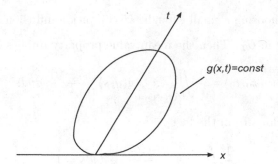

Figure 1.4: The effective region of influence.

Proof. $Pu \geq Pv$ implies $p(u - v) \geq 0$ on U_T. Thus $u - v$ has its maximum occur on the boundary G_T. By assumption $u - v \leq 0$ on G_T, and therefore $u \leq v$ in U_T. ■

In Chapter 2 we present a general comparison theorem that applies to non-linear parabolic operators.

Other consequences of maximum principles include uniqueness results. For example, we have:

Theorem 26 (Uniqueness) *Under the assumptions in Corollary 20, there can be only one solution to the problem*

$$Pu = f \quad in \ U_T,$$
$$u = g \quad on \ G_T.$$

Proof. If u and v are two solutions, then $w \equiv u - v$ satisfies $Pw = 0$ on U_T, and $w = 0$ on G_T. By Corollary 20 w attains its maximum and its minimum on G_T. Thus, $w \equiv 0$ in U_T. ■

Exercise 27 *Let $w = u - v$, where u and v are solutions to the linear convection-diffusion problem with a source*

$$u_t = \Delta u - au + q(x, t) \quad x \in \Omega \subset \mathbb{R}^3, \quad t > 0,$$
$$\nabla u \cdot n - bu = g(x, t), \quad x \in \partial\Omega, \quad t > 0,$$
$$u(x, 0) = h(x), \quad x \in \Omega,$$

where a and b are positive constants. Prove that $w = 0$ by showing

$$\frac{d}{dt} \int_\Omega w^2 dx + \frac{1}{2} \int_\Omega (\nabla w \cdot \nabla w + aw^2) dx - \frac{b}{2} \int_{\partial\Omega} w^2 dS = 0,$$

from which it follows that

$$\left(\frac{d}{dt} + const\right) \int_\Omega w^2 dx \leq 0.$$

1.6   Reference Notes

All books on elementary partial differential equations discuss the basic properties of the diffusion equation and methods for its solution. See, for example, Guenther and Lee (1996), Kevorkian (2000), Logan (1998), Strauss (1993), or Tichonov and Samarskii (1990), all of which are oriented toward applications. Intermediate level books that discuss the theory are Bassinini and Elcrat (1997), Evans (1998), John (1981), McOwen (1995), and Renardy and Rogers (1984). The latter has an excellent introduction to the functional analytic viewpoint mentioned in Section 3. Reviews by the author of recent PDE books can be found in *SIAM Review* 41(2), 393–395 (1999); 42(3), 515–522 (2000). Four texts dedicated to the diffusion equation are Cannon (1984), Carslaw and Jaeger (1959), Crank (1975), and Widder (1970). Friedmann (1964) emphasizes the general theory of parabolic equations, and Protter and Weinberger (1967) is the classical reference on maximum principles.

Chapter 2

Reaction–Advection–Dispersion Equation

A problem of great importance in environmental science is to understand how chemical or biological contaminants are transported through subsurface aquifer systems. In this chapter we consider the transport of a chemical or biological tracer carried by water through a uniform, one-dimensional, saturated, porous medium, and we derive simple mathematical models based on mass balance that incorporate advection, dispersion, and adsorption. Thus, we extend the ideas in the Chapter 1, where we focused only on the diffusion process with no flow.

The approach we take is the traditional continuum mechanics approach. Each variable, for example, density, is viewed in a mathematical sense as an idealized point function over the domain of interest. Both physically and mathematically, the values these point functions take are regarded as averages over small elementary, or representative, volume elements. The models we develop are also highly deterministic. This means they contain coefficients and functions that are regarded as completely known. In fact, the opposite may be true. There is a natural variability of nature that lends itself to an alternate approach, namely, a stochastic approach. For example, the soil conductivity, which characterizes how fast fluid can be transported through the soil fabric, is a highly variable quantity because of the ever present heterogeneities in the subsurface; one could view it as a random variable with a constant mean value over which is superimposed random spatial "noise." But this will not be the view here; this kind of random variability is not in deterministic models. Continuum-based, deterministic models include some variability, but it is disguised in through phenomenological equations obtained by the averaging process.

A complicating factor in subsurface modeling is the variability caused by the presence of several spatial scales. Aquifers can be of the order of 10^4 meters, or larger, while heterogenieties within the aquifer can range over $10^{-2} - 10^2$

29

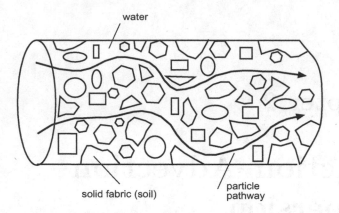

Figure 2.1: A one-dimensional porous medium showing the solid fabric, or grains, and the interstitial spaces, or pores.

meters. The pores themselves can be as small as 10^{-4} meters, while adhesive water layers, important in adsorption, may be 10^{-7} meters thick. Whether a stochastic or deterministic model is more valid is not the issue; rather, the goal is to develop a predictive model that captures the essential features of the physical processes involved.

2.1   Mass Balance

We imagine that water is flowing underground through a fixed soil or rock matrix; this soil or rock matrix, which is composed of solid material, will also be referred to as the **fabric** that makes up the porous medium. In any fixed volume, the fraction of space, called the pore space, available to the water is assumed to be ω, which is called the **porosity** of the medium. Clearly, $0 < \omega < 1$. In general, the porosity can vary with position, or even pressure, but in this chapter we assume ω is constant throughout the medium. When we say the porosity is constant, that means we are observing from a distance where there is uniformity in the porosity. If we looked on a small scale, the porosity would be either zero or one; but here we assume the representative volume element is on an order where averages of percentage pore space are constant over the entire medium. Furthermore, we assume that the flow is **saturated**, which means that all of the available pore space is always filled with fluid. There are, of course, many fuzzy issues here; for example, do we count dead end pores, where the water is trapped, as part of the pore space? Generally, no; the pore space is that space where there is mobile water. We will refer this, and other similar questions, to texts on hydrogeology [see, e.g., deMarsily (1986)]. Again, our goal is to develop simple, phenomenological mathematical models, and often some of the fine detail is omitted.

We will assume that the flow is one-dimensional, in the x-direction, and takes place in a tube of cross-sectional area A (see figure 2.1). Assuming the areal porosity is the same as the volume porosity, the cross-sectional area actually available for flow is ωA. Now let $C(x,t)$ be the concentration, measured in mass per unit volume of water, of a chemical or biological tracer dissolved in the liquid, and let $Q(x,t)$ denote its **flux**, or the rate per unit area that the contaminant mass crosses a cross section at x. We are assuming that the contaminant, for example, ions, has no volume itself and therefore does not affect the volume of the carrying liquid. We further assume that the tracer is created or destroyed with rate $F(x,t)$, measured in mass per unit volume of porous medium, per unit time. For example, F, which we call a **reaction term** or **source term**, can measure an adsorption rate, a decay rate, a rate of consumption in a chemical reaction, or even a growth or death rate if the tracer is biological. Note that the source Q can depend upon x and t through its dependence on C, i.e., $F = F(C)$. Finally, we denote by V the specific discharge, or the volume of water per unit area per unit time that flows through the medium. Note that V has velocity units, and for the present discussion we assume V is *constant*. Later we investigate the driving mechanism of the flow and set aside this constant assumption; for now we assume there is a driving mechanism that is able to maintain a constant velocity flow. We call V the **Darcy velocity**. The velocity $v \equiv V/\omega$ is called the **average velocity**; v is the velocity that would be measured by a flow meter in the porous domain. Clearly, the average velocity exceeds the Darcy velocity. Subsequently, in a constitutive equation, we shall relate the flux to the Darcy velocity. Note that we have taken the concentration to be measured in mass per unit volume. The reader should be aware that in other contexts the concentration could be measured in molarity (moles per unit volume of fluid) or molality (moles per unit mass of fluid). Recall that a mole is the mass unit equal to the molecular weight.

The basic physical law for flow in a porous medium is derived from mass balance of the chemical tracer. Mass balance states that the rate of change of the total mass in an arbitrary section of the medium must equal the net rate that mass flows into the section through its boundaries, plus the rate that mass is created, or destroyed, within the section. Therefore, consider an arbitrary section $a \leq x \leq b$ of the medium. Mass balance, written symbolically, leads directly to the integral conservation law

$$\frac{d}{dt} \int_a^b C(x,t)\omega A\,dx = Q(a,t)A - Q(b,t)A + \int_a^b F(x,t)A\,dx. \qquad (2.1)$$

The term on the left-side is the rate of change of the total amount of tracer in the section, and the first two terms on the right measure the rate that the tracer flows into the section at $x = a$ and the rate that it flows out at $x = b$; the last term is the rate that the tracer is created in the section. We assume that C and ϕ are continuously differentiable functions; thus, we may bring the time derivative under the integral sign and appeal to the fundamental theorem of calculus to write the mass balance law as

$$\int_a^b (\omega C_t + Q_x - F)\,dx = 0.$$

Because the interval of integration $[a, b]$ is arbitrary, the integrand must vanish and we obtain the mass balance law in the local, differential form

$$\omega C_t + Q_x = F. \tag{2.2}$$

Some treatments on hydrogeology measure F in units of mass per time per unit volume of water, rather than per unit volume of porous medium; then there will be an ωF term on the right-side of (2.2).

At this point, a constitutive relation, usually based in empirics, must be postulated regarding the form of the flux Q. We should ask how dissolved particles get from one place to another in a porous medium. There are three generally accepted ways. One way is by **advection**, which means that particles are simply carried by the bulk motion of the fluid. This leads us to define the **advective flux** $Q^{(a)}$ by

$$Q^{(a)} = VC,$$

which is just the product of the velocity and concentration. Another method of transport is by **molecular diffusion**. This is the spreading caused by the random molecular motion and collisions of the particles themselves. This is precisely the type of diffusion discussed in Chapter 1; there we stated that this type of motion is driven by concentration gradients and the flux due to diffusion is given by Fick's law. We call this the **molecular diffusion flux** $Q^{(m)}$ and we take

$$Q^{(m)} = -\omega D^{(m)} C_x.$$

$D^{(m)}$ is the **effective molecular diffusion coefficient** in the porous medium. The diffusion occurs in a liquid phase enclosed by the solid porous fabric. The solid boundaries hinder the diffusion, and therefore $D^{(m)}$ is smaller than the usual molecular diffusion coefficient D_0 that one would measure in an immobile liquid with no solid boundaries. The reduction in the diffusion coefficient is therefore caused by the structure of the porous fabric and the presence of the tortuous flow paths available to the fluid. The ratio $D^{(m)}/D_0$ is often called the **tortuosity** of the medium and varies roughly from 0.1 to 0.7. Molecular diffusion is present whether or not the fluid is moving.

There is a third contribution to the particle flux called **kinematic** (or, **mechanical) dispersion**. This is the spreading, or mixing phenomenon, caused by the variability of the complex, microscopic velocities through the pores in the medium. So, it is linked to the heterogeneities present in the medium and is present only if there is flow. The idea is that different flow pathways have different velocities and some have a greater than average velocity to carry the solutes ahead of a position based only on the mean velocity. The mathematical form of the dispersion flux $\phi^{(d)}$ is taken to be Fickian and given by

$$Q^{(d)} = -\omega D^{(d)} C_x,$$

where $D^{(d)}$ is the **dispersion coefficient**. Thus, the net flux is given by the sum of the advective, molecular, and dispersion fluxes:

$$\begin{aligned} Q &= Q^{(a)} + Q^{(m)} + Q^{(d)} \\ &= VC - \omega(D^{(m)} + D^{(d)})C_x. \end{aligned}$$

If we define the **hydrodynamic dispersion coefficient** D by

$$D = D^{(m)} + D^{(d)},$$

then the net flux is given simply as

$$Q = VC - \omega DC_x.$$

The Fickian term $-\omega DC_x$ is termed the **hydrodynamic dispersion**. It consists of molecular diffusion and kinematic dispersion.

To summarize, when there is no flow velocity, the only flux is molecular diffusion. When there is flow, we get advection and dispersion as well. So, if there is a "plume" of contaminant in the subsurface, we expect it to advect with the bulk motion of the fluid and spread out from diffusion and dispersion. Thus, dispersion adds a spreading effect to the diffusional effects. Generally, it is observed in three dimensions that the spreading caused by dipersion is greater in the direction of the flow than in the transverse directions. If no dispersion were present, a spherical plume would just spread uniformly as it advected with the flow. This means that in a higher-dimensional formulation of the equations, the hydrodynamic dispersion would be different in different directions.

Because dispersion is present in moving fluids, it has been an important exercise to determine how the dispersion coefficient depends on the velocity of the flow. Experiments have identified several flow regimes where different mechanisms dominate. These flow regimes are usually characterized by the Peclet number

$$Pe = \frac{|V|l}{\omega D_0},$$

where l is an intrinsic length scale, say the mean diameter of pores. For very low velocity flows, i.e., very small Peclet numbers, molecular diffusion dominates dispersion. As the Peclet number increases, both processes are comparable until dispersion begins to dominate and diffusion becomes negligible; this occurs for approximately $Pe > 10$. Generally, for many velocities of interest, experimenters propose, in the direction of the flow, the linear constitutive relationship

$$D^{(d)} = \alpha_L |V|,$$

where α_L is the **longitudinal dispersivity**. In transverse directions to the flow, the dispersion coefficient is taken to be $\alpha_T |V|$, where the transverse dispersivity α_T is roughly an order of magnitude smaller that the longitudinal dispersivity. If we make these assumptions, the hydrodynamic dispersion coefficient in one-dimensional flow can be written

$$D = D^{(m)} + \alpha_L |V|.$$

When the velocity is small, the dispersion is negligible, and when the velocity is large, the dispersion will dominate the diffusion. Details of experiments and numerical values of the Peclet number ranges and dispersivities can be found in hydrogeology texts [see deMarsily (1986) or Domenico and Schwartz (1990)].

Combining the constitutive relations with the mass balance law (2.2) gives the fundamental **reaction-advection-dispersion equation**

$$\omega C_t = (\omega D C_x)_x - V C_x + F. \tag{2.3}$$

If D is constant, then D can be pulled out of the derivative and we can write

$$C_t = D C_{xx} - v C_x + \omega^{-1} F.$$

We remark that geologists, civil engineers, mathematicians, and so on, frequently use different terminology in describing the phenomena embodied in equation (1.3). Thus, advection is often termed convection, and dispersion is replaced with diffusion. The reaction term F is a source. Therefore, equation (2.3) is sometimes called a reaction-convection-diffusion equation, or a convection-diffusion equation with sources.

Observe that special cases of equation (2.3) are the **dispersion (or diffusion) equation,**

$$C_t = D C_{xx},$$

which was the subject of Chapter 1, and the simple **advection equation**

$$C_t = -v C_x.$$

The advection equation has a general solution of the form of a **right-traveling wave**

$$C(x,t) = U(x - ct),$$

where U is an arbitrary function. These types of solutions are the subject of Chapter 3.

Example 28 *If the tracer is radioactive with decay rate λ, then $F = -\lambda \omega C$ and we obtain the linear* **advection-dispersion-decay** *equation*

$$C_t = D C_{xx} - v C_x - \lambda C.$$

A change of dependent variable to $w = Ce^{\lambda t}$ leads to an equation without the decay term, and a transformation of independent variables to $\tau = t$, $\xi = x - vt$ eliminates the advection term. Hence, the advection-dispersion-decay equation can be transformed into a simple diffusion equation. The complete transformation is

$$c(x,t) = w(x,t)e^{\frac{v}{2D}(x-vt)-\lambda t},$$

which gives $w_t = D w_{xx}$.

Example 29 *If the tracer is a biological species with* **logistic growth** *rate* $F = rC(1 - C/K)$, *where* r *is the growth constant and* K *is the carrying capacity, then*

$$C_t = DC_{xx} - vC_x + \frac{r}{\omega}C\left(1 - \frac{C}{K}\right),$$

which is an advection-dispersion equation with growth.

2.2 Several Dimensions

Now let us consider a porous domain Ω in $\mathbb{R}^3$. Before stating the mass balance law we briefly review some notation. Points in $\mathbb{R}^3$ are denoted by $x = (x_1, x_2, x_3)$ or just (x, y, z). Because of the context, there should be no confusion in using x sometimes as a point and other times as a coordinate. A volume element is $dx = dx_1 dx_2 dx_3$. Volume integrals over Ω are denoted by

$$\int_\Omega f(x)dx,$$

and flux integrals over the surface $\partial\Omega$ of Ω are denoted by

$$\int_{\partial\Omega} Q \cdot n dA.$$

Here, f is a scalar function, Q is a vector function,[1] and n is the outward unit normal. Note that we are dispensing with writing triple and double integrals.

If we apply mass balance to an arbitrary ball (spherical region) Ω' in Ω, the integral conservation law takes the form

$$\frac{d}{dt}\int_{\Omega'} \omega C dx = -\int_{\partial\Omega'} Q \cdot n dA + \int_{\Omega'} F dx, \tag{2.4}$$

where $\partial\Omega'$ denotes the surface of the ball Ω'. Here, n is the outward unit normal vector the concentration is $C = C(x, t)$, where $x \in \mathbb{R}^3$, and the flux vector is Q. As in one dimension, equation (2.4) states that the rate of change of solute in the ball equals the net flux of solute through the surface of the ball plus the rate that solute is produced in the ball. The divergence theorem allows us to rewrite the surface integral and (2.4) becomes

$$\frac{d}{dt}\int_{\Omega'} \omega C dx = -\int_{\Omega'} \nabla \cdot Q\, dx + \int_{\Omega'} F dx.$$

Owing to the arbitrariness of the domain Ω', we obtain the local form of the conservation law as

$$(\omega C)_t = -\nabla \cdot Q + F, \quad x \in \Omega.$$

---

[1] We shall not use special notation for vectors; whether a quantity is a vector or scalar will be clear from the context.

Here we are assuming that the functions are sufficiently smooth to allow application of the divergence theorem and permit pulling the time derivative inside the integral. As in the one-dimensional case we have too many unknowns, and so additional equations, in the form of constitutive relations, are needed. In particular, we assume that the vector flux is made up of a dispersive-diffusion term and an advective term and is thus related to the concentration via

$$Q = -\omega D \nabla C + CV,$$

where V is the vector Darcy velocity and D is the hydrodynamic dispersion coefficient. Then the mass balance law becomes

$$\omega C_t = \nabla \cdot (\omega D \nabla C) - \nabla \cdot (CV) + F.$$

If the flow is incompressible, then $\nabla \cdot V = 0$ and we obtain

$$\omega C_t = \nabla \cdot (\omega D \nabla C) - V \cdot \nabla C + F, \tag{2.5}$$

which is the **reaction–advection–dispersion equation** in three dimensions. If the dispersion coefficient D is a pure constant, then it may be brought out of the divergence and we obtain the equation in the form

$$C_t = D \Delta C - v \cdot \nabla C + \omega^{-1} F,$$

where Δ is the three-dimensional Laplacian operator (recall the vector identity $\Delta C = \nabla \cdot \nabla C$) and $v = V/\omega$ is the average velocity vector.

In general, the dispersion coefficient D in (2.5) is not constant; in fact, usually $D = D(x, C)$, which means that it may depend on position if the medium is nonhomogeneous or even the concentration. Furthermore, it may have different values in different directions if the medium is not isotropic. In the special case that the hydrodynamic dispersion coefficient varies in the coordinate directions we have

$$D = \text{diag}(D^{(x)}, D^{(y)}, D^{(z)}),$$

which is a diagonal matrix. Then the mass balance law is expanded to

$$\omega C_t = (\omega D^{(x)} C_x)_x + (\omega D^{(y)} C_y)_y + (\omega D^{(z)} C_z)_z - V \cdot \nabla C + F.$$

In many hydrogeological applications the flow field and geometry lends itself to a description in cylindrical coordinates. This occurs, for example, in the pumping of wells and boreholes. In cylindrical geometry the mass balance equation (2.5) can be written as

$$C_t = \frac{1}{r}(r D^{(r)} C_r)_r + \frac{1}{r^2}(D^{(\theta)} C_\theta)_\theta + (D^{(z)} C_z)_z - \nabla \cdot (Cv) + \omega^{-1} F,$$

where we have used the fact that the gradient operator is

$$\nabla = (\partial_r, \frac{1}{r}\partial_\theta, \partial_z),$$

and the average velocity is $v = (v^{(r)}, v^{(\theta)}, v^{(z)})$ in cylindrical coordinates. The quantities $D^{(r)}, D^{(\theta)}, D^{(z)}$ represent the hydrodynamic dispersion coefficients in the coordinate directions. [Expressions for the divergence, gradient, and Laplacian in various coordinate systems can be found in many calculus texts, texts in fluid mechanics or electromagnetic theory. A good reference is Bird, Stewart, and Lightfoot (1960). See also Sun (1995) for a general formulation in orthogonal curvilinear coordinates].

An important case of this latter equation occurs when the velocity field is radial, i.e., $v = (v^{(r)}, 0, 0)$, and there is negligible dispersion in the vertical or angular directions, i.e., $D^{(z)} = D^{(\theta)} = 0$. Then we obtain the radial reaction-advection-dispersion equation

$$C_t = \frac{1}{r}(rD^{(r)}C_r)_r - v^{(r)}C_r - \frac{C}{r}(rv^{(r)})_r + \omega^{-1}F.$$

Specifically, if the radial velocity is

$$v^{(r)} = \frac{a}{r},$$

then the velocity field is divergence-free and the radial equation becomes

$$C_t = \frac{1}{r}(rD^{(r)}C_r)_r - \frac{a}{r}C_r + \omega^{-1}F.$$

Using the constitutive assumption that the kinematic dispersion coefficient is proportional to the magnitude of the velocity and diffusion is negligible, that is, $D^{(r)} = \alpha_r ||v|| = a\alpha_r/r$, we have

$$C_t = \frac{a\alpha_r}{r}C_{rr} - \frac{a}{r}C_r + \omega^{-1}F. \tag{2.6}$$

We can think of equation (2.6) as modeling the transport of a contaminant in a radial flow field. If $a < 0$, then the flow is toward an extraction well and the equation models a **remediation process**; if $a > 0$, then the flow is radially outward from a well and the process is a **contamination process**. The value of a depends on the pumping rate. In Section 2.5 we solve a special radial dispersion problem.

Boundary conditions on the concentration are also a necessary ingredient in model formulation. We discuss different types of boundary conditions in Section 2.7.

2.3 Adsorption Kinetics

Several geochemical mechanisms can change the character of transported substances through porous domains. One such important mechanism is **adsorption**. Adsorption is a process, often brought on by ion exchange, that causes the mobile tracer, or solute, to adhere to the surface of the solid porous fabric, and thus become immobile. To model such processes in one dimension we let

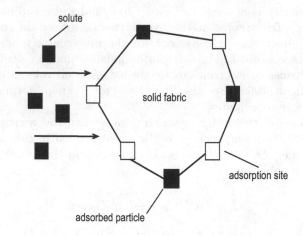

Figure 2.2: Solute-site-adsorbed particle reaction.

$S = S(x,t)$ denote the amount of solute adsorbed. Because the solutes become attached to mineral particles, this amount S of solute adsorbed is usually measured in mass of solute per unit mass of soil. The mass of soil per unit volume of porous medium is $\rho(1 - \omega)$ and therefore the rate of adsorption is given by

$$F = -\rho(1 - \omega)S_t.$$

Consequently, under the assumption of constant D, the mass balance equation (2.3) becomes

$$C_t = DC_{xx} - vC_x - \frac{\rho(1 - \omega)}{\omega}S_t, \qquad (2.7)$$

which is the **adsorption-advection-dispersion** equation. In Chapter 5 we examine different kinds of geochemical reactions, namely, those that change the mineralogy of the host rock and even change the porosity of medium.

2.3.1   Instantaneous Kinetics

In the simplest case we can envision the adsorption process as a reversible chemical reaction where one adsorption site on the solid reacts with a solute particle to produce an absorbed particle. The reversibility of the reaction means that, in a strict sense, the process is an adsorption-desorption process. A schematic is shown in figure 2.2, and we represent the reaction as

$$[\sigma] + [C] \rightleftharpoons [S].$$

Here σ denotes the density of adsorption sites on the immobile solid fabric.

If the number of adsorption sites is large, then we may take σ to be constant. Then the law of mass action states that the reaction rate is

$$r = k_f \sigma C - k_b S,$$

where k_f and k_b are the forward and backward rate constants, respectively. If we assume that the reaction equilibrates on a fast time scale compared to that of dispersion and advection, then the reaction is always in equilibrium ($r \equiv 0$), or

$$S = K_d C, \qquad (2.8)$$

where $K_d \equiv k_f \sigma / k_b$. Thus, there is an instantaneous, algebraic, linear relation between the concentration of solute and the concentration of adsorbed particles. Equation (2.8) is called a **linear adsorption isotherm** and K_d is the **distribution constant** (a typical value is $K_d = 0.476\ \mu\text{gm/gm}$). Substituting (2.8) into (2.7) yields, after rearrangement, an advection–dispersion equation

$$R_f C_t = D C_{xx} - v C_x, \qquad (2.9)$$

where R_f is the **retardation constant** given by

$$R_f = 1 + \frac{\rho K_d (1 - \omega)}{\omega}.$$

Thus, a linear adsorption process reduces the apparent speed of advection to v / R_f. The literature in hydrogeology is full of closed-form, or analytic, solutions to this equation with various initial and boundary data.

If there is a limited number of adsorption sites with density σ_0, then

$$S + \sigma = \sigma_0,$$

and the equilibrium condition is $R = k_f (\sigma_0 - S) C - k_b S = 0$, or

$$S = \frac{k_f \sigma_0 C}{k_f C + k_b}. \qquad (2.10)$$

This relationship is the **Langmuir adsorption isotherm**. Now the mass balance equation (2.7) is nonlinear and given by

$$\left(C + \frac{\rho (1 - \omega)}{\omega} \frac{k_f \sigma_0 C}{k_f C + k_b} \right)_t = D C_{xx} - v C_x. \qquad (2.11)$$

In general, if the reaction equilibrates on a time scale that is fast compared to that of advection and dispersion, then we assume an algebraic relationship between the solute concentration C and the adsorbed concentration S. Such relations, which hold in an equilibrium state, are said to be instantaneous and define the adsorption kinetics. The algebraic relationship, called the **adsorption isotherm**, often takes the form

$$S = f(C), \qquad (2.12)$$

where the function f usually has the properties:

$$f \in C^2(0, \infty); \quad f(0) = 0; \quad f'(C) > 0, \quad f''(C) < 0 \text{ for } C > 0. \tag{2.13}$$

For some isotherms $f'(0)$ may not exist, as in the case of the Freundlich isotherm below.

In addition to the linear and the Langmuir isotherms, (2.8) and (2.10), respectively, other adsorption isotherms are suggested in the literature. A partial list is the following:

$$S = K_d C, \qquad\qquad \text{(linear)}$$

$$S = \frac{k_f \sigma_0 C}{k_f C + k_b} \qquad\qquad \text{(Langmuir)}$$

$$S = kC^{1/n}, \quad n > 1 \qquad\qquad \text{(Freundlich)}$$

$$S = k_1 C - k_2 C^2 \qquad\qquad \text{(quadratic)}$$

$$S = \frac{k_1 C^m}{1 + k_2 C^m} \qquad\qquad \text{(generalized Langmuir)}$$

$$S = k_1 e^{-k_2/C} \qquad\qquad \text{(exponential)}$$

The Freundlich isotherm is one widely applied to the adsorption of various metals and organic compounds in soils. Unfortunately there is no upper limit to the amount of solute that can be adsorbed, so the Freundlich isotherm must be used with caution in experimental studies.

In summary, with kinetics given by (2.12), the mass balance equation takes the form

$$(C + \beta f(C))_t = DC_{xx} - vC_x,$$

or, equivalently,

$$(1 + \beta f'(C))C_t = DC_{xx} - vC_x, \tag{2.14}$$

where $\beta = \rho(1 - \omega)/\omega$ and f satisfies the conditions (2.13).

2.3.2   Noninstantaneous Kinetics

If the reaction proceeds so slowly that it does not have time to come to local chemical equilibrium, then the kinetics of adsorption–desorption is not instantaneous and reiquires a dynamic rate law for its description. Such a law has the form

$$S_t = F(S, C). \tag{2.15}$$

So the rate of adsorption depends on both C and S. We could reason, for example, that the adsorption rate should increase as the concentration C of solute increases ($F_C \geq 0$); but, as more and more chemical is adsorbed, the ability of the solid to adsorb additional chemical will decrease ($F_S \leq 0$). The simplest such model with these characteristics is

$$S_t = F(S, C) = k_1 C - k_2 S, \quad k_1, k_2 > 0.$$

For nonequilibrium adsorption–desorption, solute-soil reactions, the coupled pair of equations

$$C_t = DC_{xx} - vC_x - \frac{\rho(1-\omega)}{\omega}S_t, \tag{2.16}$$

$$S_t = F(S,C) \tag{2.17}$$

form a general model. We shall always assume that

$$F_C \geq 0, \quad F_S \leq 0. \tag{2.18}$$

Exercise 30 *For the linear system*

$$C_t = DC_{xx} - vC_x - \beta S_t, \quad \beta \equiv \frac{\rho(1-\omega)}{\omega} \tag{2.19}$$

$$S_t = k_1 C - k_2 S, \tag{2.20}$$

eliminate S to obtain a single equation for C by the following scheme. Substitute (2.20) into (2.19) and then take the time derivative to get

$$C_{tt} = DC_{xxt} - vC_{xt} - \beta(k_1 C_t - k_2 S_t).$$

Then multiply (2.19) by k_2 and subtract the result from the last equation to obtain

$$C_{tt} + (\beta k_1 - k_2)C_t = DC_{xxt} - k_2 DC_{xx} - vC_{xt} + vk_2 C_x.$$

This is a third-order differential equation for the concentration C. It can be compared with the wave equation with internal damping

$$u_{tt} = c^2 u_{xx} + a u_{xxt}$$

[see, for example, Guenther and Lee (1991)].

Example 31 *In the case that the transported particles are colloids, Saiers, et al. (1994) have given a kinetics law of the form*

$$S_t = \frac{\omega K_d}{\rho_b}C\frac{a-S}{a} - kS,$$

where ρ_b is the bulk density of the solid fabric, k is the entrainment coefficient, a is the colloidal retention capacity. [A mathematical analysis of this model can be found in Cohn and Logan (1995)].

2.3.3 Mulitple-Site Kinetics

In some cases attachment of the solute to soil particles can occur in different ways with different kinetics. For example, such differences can arise because of different adsorption behavior of planar sites and edge sites on the soil. Or, different adsorption sites may have different accessibility. Let us assume there

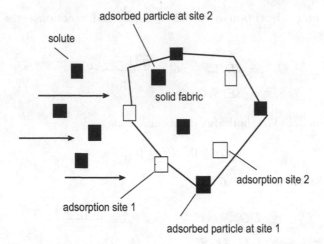

Figure 2.3: Solute-particle reaction for multiple sites represented schematically by interior and surface locations.

are two ensembles of adsorption sites represented by densities σ^1 and σ^2. See figure 2.3.

Let $S^1 = S^1(x,t)$ and $S^2 = S^2(x,t)$ denote the concentrations of adsorbed material at sites 1 and 2, respectively. The reaction can be represented symbolically by

$$
\begin{aligned}
[\sigma^1] + [C] &\rightleftharpoons [S^1], \\
[\sigma^2] + [C] &\rightleftharpoons [S^2].
\end{aligned}
$$

The net rate of adsorption is $S_t = (S^1 + S^2)_t$ and therefore the adsorption–advection–dispersion equation becomes

$$
C_t = DC_{xx} - vC_x - \frac{\rho(1-\omega)}{\omega}(S_t^1 + S_t^2).
$$

In the case both reactions are in local chemical equilibrium we supplement this equation with the two isothermal relations

$$
S^1 = f_1(C), \quad S^2 = f_2(C).
$$

If both are slow to equilibrate, then we have

$$
S_t^1 = F_1(C, S^1), \quad S_t^2 = F_2(C, S^2).
$$

In the mixed case where the first reaction equilibrates rapidly and the second equilibrates slowly, we have

$$
S^1 = f_1(C), \quad S_t^2 = F_2(C, S^2).
$$

2.4 Dimensionless Equations

Generally, in applied mathematics we study equations in their dimensionless form. Not only does the dimensionless model usually contain fewer parameters, but in applying perturbation methods to obtain approximations it is essential to scale the problem properly so that small parameters are correctly place in the governing equations [see Lin and Segel (1989) or Logan (1997c) for a general discussion of the importance of scaling and dimensional analysis]. If L is some length scale for the problem, then, unless otherwise noted, we select the time scale to be L/v, which is the **advection time scale**. Other time scales can be chosen, for example, one based on diffusion or one based on the reaction rate. Concentrations can be scaled by some reference concentration C_0, which could be the maximum initial or boundary concentration at an inlet. Therefore, we introduce dimensionless space, time, and concentration variables via

$$\xi = \frac{x}{L}, \quad \tau = \frac{t}{L/v}, \quad u = \frac{C}{C_0}.$$

Then the reaction–advection–dispersion model

$$C_t = DC_{xx} - vC_x + \omega^{-1}F$$

becomes

$$u_\tau = \alpha u_{\xi\xi} - u_\xi + f,$$

where

$$\alpha \equiv Pe^{-1} = \frac{D}{vL}, \quad q \equiv \frac{L}{\omega v C_0}F$$

are dimensionless quantities. Here, α is the reciprocal of what is called the **Peclet number**, which measures the ratio of advection to dispersion; q is the source term and u is the dimensionless concentration. Usually, we will just use x and t for the dimensionless spatial and time variables in place of ξ and τ and write the **reaction–advection–dispersion model** as

$$u_t = \alpha u_{xx} - u_x + q. \tag{2.21}$$

In the same manner, we will write the **equilibrium model** (2.7) and (2.12) in dimensionless form as

$$u_t = \alpha u_{xx} - u_x - \beta s_t, \tag{2.22}$$
$$s = f(u). \tag{2.23}$$

Here s is the dimensionless adsorbed concentration (scaled by C_0) and $\beta > 0$ is a dimensionless constant. The dimensionless **nonequilibrium model** (2.16)–(2.17) is

$$u_t = \alpha u_{xx} - u_x - \beta s_t, \tag{2.24}$$
$$s_t = F(u, s). \tag{2.25}$$

Observe that the equilibrium equation is of the form

$$g(u)_t = \alpha u_{xx} - u_x,$$

where $g(u) = u + \beta f(u)$. Because $g'(u) > 0$, we may define the one-to-one transformation $w = g(u)$. Then, if $u = h(w)$ is the inverse transformation, we have $u_x = h'(w)w_x$ and the equation may be written

$$w_t = \alpha(h'(w)w_x)_x - h'(w)w_x.$$

Thus we have succeeded in transforming the equilibrium equation to a uniformly parabolic equation.

Much is known about nonlinear **reaction–diffusion equations**, without advection, of the form

$$u_t = (k(u)u_x)_x + F(u).$$

For example, they are discussed in detail in Samarski *et al.* (1995). Such nonlinear models are commonplace in population dynamics and in combustion theory.

2.5   Nonlinear Equations

In the preceding sections we developed some of the basic nonlinear mathematical models in contaminant transport. We now illustrate, by way of examples, some of the elementary properties of nonlinear equations and techniques used to study them. These techniques include comparison methods, similarity methods, and energy methods. Later, in Chapter 3, we study traveling waves.

The partition of differential equations into linear and nonlinear models is a significant one. Linear equations are sometimes solvable; in any case, they have a form that permits the vast tools of linear analysis, or functional analysis, to be applied to determine their behavior and solution structure. Nonlinear equations are nearly always unsolvable by analytic methods, and to obtain specific solutions we nearly always resort to numerical methods. Generally, the tools of nonlinear analysis are not so nearly well-developed and all-encompassing as in linear analysis. Nevertheless, there are some general principles and techniques that are available that take us beyond just *ad hoc* methods.

We emphasize again that we are studying model equations. Using the term "model" helps us realize the distinction between reality and theory. Models do not include all of the details of the physical reality. In the best of all possible worlds, the model should give a reasonable description of some part of reality. This is why we often separate out the mechanisms and study equations only with diffusion or only with advection. Understanding the behavior of these simple models can then give us clues into the behavior of more general problems. For example, if we can show that some simple, model, nonlinear reaction–advection–dispersion equation has solutions that blow up (go to infinity, or have their derivatives go to infinity) at a finite time, then we have succeeded in creating a healthy skepticism about such equations. Therefore, when we as applied scientists or mathematicians develop detailed descriptions of other, more complicated, systems, we will have insights into their behavior and may not unconscientiously believe that our model has solutions that exist for all time. It is always tempting (and dangerous!) to believe that our descriptions of reality automatically lead to well-posed problems.

2.5.1 A Comparison Principle

In Section 1.5 we stated and proved a comparison principle for a linear parabolic equation. Such results extend to nonlinear equations. We begin with some terminology. Consider the partial differential equation

$$u_t = F(x, t, u, u_x, u_{xx}). \tag{2.26}$$

For notational simplicity, we denote $p = u_x, r = u_{xx}$, and we write the function F as $F = F(x, t, u, p, q)$. We assume that F is a continuously differentiable function of its five variables. We say the function F is **elliptic** with respect to a function $u = u(x, t)$ at a point (x, t) if

$$F_r(x, t, u, p, r) > 0 \quad \text{at} \quad (x, t). \tag{2.27}$$

The function F is elliptic in a domain D of space time if it is elliptic at each point of D. If F is elliptic in D, then we say that equation (2.26) is **parabolic** in D. We say the function F is **uniformly elliptic** with respect to a function $u = u(x, t)$ in D if F_r is bounded away from zero; that is, there exists a positive constant μ such that

$$F_r(x, t, u, p, r) \geq \mu > 0 \quad \text{for all } (x, t) \in D. \tag{2.28}$$

In this case we say that equation (2.26) is **uniformly parabolic** in D. Although D may be quite general, we usually take $D = I \times (0, T)$, where I is an open interval in $\mathbb{R}$ (possibly unbounded).

For example, if $D(u)$ is a nonnegative, continuously differentiable function and $H(u)$ is continuously differentiable, then the nonlinear advection–dispersion equation

$$u_t = (D(u)u_x)_x + H(u)_x \tag{2.29}$$

is parabolic for all functions u; the parabolicity condition (2.27) becomes simply $D(u) > 0$. If the dispersion coefficient is bounded away from zero, that is, $D(u) \geq d_0 > 0$, then it is uniformly parabolic; in this case we say the equation (2.26) is **nondegenerate**. If the diffusion coefficient satisfies only the condition $D(u) > 0$, then we say the advection–dispersion equation (2.28) is **degenerate**.

Another example is the equilibrium model

$$g(u)_t = \alpha u_{xx} - u_x, \quad g(u) \equiv u + \beta f(u), \tag{2.30}$$

where the isotherm is positive, increasing, and concave downward, and $f(0) = 0$. As noted in Section 2.2, we can transform this equation to standard parabolic form by letting $w = g(u)$. This mapping is one-to-one and invertible; we denote the inverse transformation by $u = h(w)$. Then the model (2.28) becomes

$$w_t = (\alpha h'(w)w_x)_x - h'(x)w_x,$$

where

$$h'(w) = \frac{1}{g'(u)} = \frac{1}{1 + \beta f'(u)}, \qquad u = h(w).$$

Clearly, if $f'(0^+) = +\infty$, then $h'(0^+) = 0$ and the equation is degenerate.

Exercise 32 *In the case of a Freundlich isotherm $f(u) = \sqrt{u}$, we have $f'(0^+) = 0$ and equation (2.30) is degenerate. Show that*

$$h(w) = (\sqrt{(\beta/2)^2 + w} - \beta/2)^2,$$

which gives

$$h'(w) = \frac{\sqrt{\beta^2 + 4w} - \beta}{\sqrt{\beta^2 + 4w}}.$$

Note that $h'(0) = 0$.

Exercise 33 *In the case of a Langmuir isotherm $f(u) = \frac{u}{1+u}$, show that*

$$h'(w) = \frac{(1 + u)^2}{\beta + (1 + u)^2}.$$

Check that $h'(v) \geq 1/(1 + \beta)$, so the equation (2.30) is nondegenerate.

Nonlinear parabolic equations of the form (2.26) admit what is called a **comparison principle,** which allows us to compare solutions to similar problems, say, differing only in their initial or boundary data. If one of the problems can be solved, then a comparison principle can give bounds on the solution to the problem that perhaps cannot be solved. This information can be used, for example, to prove positivity of solutions, to obtain asymptotic estimates of the behavior of solutions for large times, or produce a priori bounds that guarantee existence of solutions.

The basic comparison result can be stated as follows [see Protter and Weinberger (1967)]:

Theorem 34 *Let I be a bounded spatial interval in $\mathbb{R}$ and let $D = I \times (0, T]$, and let L denote the operator defined by*

$$Lu \equiv u_t - F(x, t, u, u_x, u_{xx}).$$

Suppose that u, w, and W are continuous on the closure of D and twice continuously differentiable on D. Furthermore, assume that F is elliptic on D with respect to the functions $\theta w + (1 - \theta)u$ and $\theta W + (1 - \theta)u$ for all $0 \leq \theta \leq 1$. If

$$Lw \leq Lu \leq LW \quad \text{in } D,$$

and

$$w(x,0) \leq u(x,0) \leq W(x,0) \quad x \in I,$$

and

$$w(x,t) \leq u(x,t) \leq W(x,t) \quad x \in \partial I, \ 0 \leq t \leq T,$$

then

$$w(x,t) \leq u(x,t) \leq W(x,t) \quad in \ D.$$

Example 35 *Consider the initial boundary value problem*

$$
\begin{align}
u_t &= (D(u)u_x)_x - u_x, \ x \in (a,b), \ t > 0, \tag{2.31}\\
u(x,0) &= u_0(x), \ x \in (a,b), \tag{2.32}\\
u(a,t) &= u_1(t), \ u(b,t) = u_2(t), \ t > 0, \tag{2.33}
\end{align}
$$

where u_1, u_2, and u_0 are nonnegative. Taking w to be the identically zero function $w \equiv 0$ shows that $u(x,t) \geq 0$, giving a positivity result for a classical solution $u(x,t)$.

Exercise 36 *Consider the equilibrium model*

$$
\begin{align}
g(u)_t &= \alpha u_{xx} - u_x, \ x \in (a,b), \ t > 0, \tag{2.34}\\
u(x,0) &= u_0(x), \ x \in (a,b), \tag{2.35}\\
u(a,t) &= u_1(t), \ u(b,t) = u_2(t), \ t > 0, \tag{2.36}
\end{align}
$$

where $g(u) \equiv u + \beta f(u)$ and where u_0, u_1, and u_2 are nonnegative. Show that this problem can be transformed into

$$
\begin{align}
w_t &= (\alpha h'(w)w_x)_x - h'(x)w_x, \ x \in (a,b), \ t > 0, \tag{2.37}\\
w(x,0) &= w_0(x), \ x \in (a,b), \tag{2.38}\\
w(a,t) &= ww_1(t), \ w(b,t) = w_2(t), \ t > 0, \tag{2.39}
\end{align}
$$

where $w_0 = u_0 + \beta f(u_0) \geq 0, w_1 = u_1 + \beta f(u_1) \geq 0, w_2 = u_2 + \beta f(u_2) \geq 0$. Use the comparison principle to show that $w(x,t)$, and hence $u(x,t)$, is nonnegative.

The comparison principle can also be applied to prove a maximum principle for the equilibrium model (2.37)–(2.39). Let $m > 0$ be the maximum of the data $\{w_0, w_1, w_2\}$ for $x \in [a,b]$ or $t \in [0,T]$. Then, if w is a classical solution to (2.37)–(2.39), then

$$Lw \equiv w_t - (\alpha h'(w)w_x)_x + h'(x)w_x = 0 \leq L(m) = 0.$$

This implies $w(x,t) \leq m$ in $[a,b] \times [0,T]$.

2.5.2   Similarity Solutions

In Chapter 1 we observed that the classical, linear diffusion equation has solutions that are smooth, even when the boundary and initial data are discontinuous. The diffusion equation smooths out solutions. When the diffusion is
nonlinear, however, this smoothing property no longer holds; nonlinear diffusion
or dispersion equations can propagate piecewise smooth solutions in much the
same manner as a wave-like equation. This type of phenomenon occurs when
the equation is degenerate, i.e., when the dispersion coefficient $D = D(u)$ can
vanish. Intuitively, one can think physically that when the dispersion coefficient vanishes, then dispersion is not present to smooth out irrregularities in
the solution.

We illustrate this type of behavior using a standard technique for obtaining solutions to nonlinear parabolic problems on unbounded domains. Let us
consider the equilibrium model on a semi-infinite domain with no advection:

$$
(u + \beta f(u))_t \;=\; \alpha u_{xx}, \quad x > 0,\ t > 0, \tag{2.40}
$$

$$
u(0, t) \;=\; 1, \quad t > 0, \tag{2.41}
$$

$$
u(x, 0) \;=\; 0, \quad x > 0, \tag{2.42}
$$

$$
u(x, t) \;\to\; 0 \ as \ x \to \infty, \tag{2.43}
$$

where f satisfies the conditions $f(0) = 0, f \in C^1[0, \infty) \cup C^2(0, \infty)$, with $f >
0,\ f' > 0,\ f'' < 0$. The similarity method is a general method to construct
solutions when the differential equation is invariant under a local Lie group of
transformations. These internal symmetries lead to classes of invariant solutions called **similarity solutions**. We refer the reader to one of the following
references for a general discussion of the similiarity method: Dresner (1983),
Logan (1987, 1994). Here, motivated by the form of solutions to the classical,
linear heat equation, we shall attempt a solution of the form

$$
u(x, t) = y(s),
$$

where the similarity variable s is given by

$$
s = \frac{x}{\sqrt{t}}.
$$

Substituting the expresson for u into the PDE gives an ordinary differential
equation for the function y, namely,

$$
-\frac{s}{2}\frac{d}{ds}(y + \beta f(y)) = \alpha \frac{d^2 y}{ds^2}.
$$

The initial and boundary conditions force

$$
y(0) = 1, \quad y(+\infty) = 0.
$$

Integrating the differential equation with respect to s and using the boundary
condition at $s = +\infty$ to evaluate the constant of integration yields

$$
\frac{dy}{ds} = -\frac{s}{2\alpha}(y + \beta f(y)).
$$

Separating variables and integrating gives the formal, implicit solution to (2.40)–(2.43) as

$$\int_1^y \frac{dw}{w + \beta f(w)} = -\frac{s^2}{4\alpha},$$

or, in original variables,

$$\int_1^u \frac{dw}{w + \beta f(w)} = -\frac{x^2}{4\alpha t}.$$

Exercise 37 *As an example of the preceding calculation, verify the details of the following special case. Consider the case of a Langmuir isotherm where*

$$f(u) = \frac{k_1 u}{k_2 + u}.$$

A partial fractions expansion gives

$$\int \frac{dw}{w + \beta f(w)} = \ln(w^a (w + A)^b),$$

where $a = k_2/A$, $b = 1 - k_2/A$, $A = \beta k_1 + k_2 > 0$. *Moreover,* $0 < a < 1$ *and* $b > 0$. *Therefore,*

$$\ln(w^a (w + A)^b) \,|_1^y = -\frac{s^2}{4\alpha},$$

or

$$s^2 = -4\alpha \ln \left(y^a \left(\frac{y + A}{1 + A} \right)^b \right).$$

This equation defines, implicitly, a solution $u(x,t) = y(x/\sqrt{t})$.

To illustrate another property of nonlinear diffusion equations, consider

$$u_t = (u u_x)_x, \quad x \in \mathbb{R}, \ t > 0. \tag{2.44}$$

This is a special case of the **porous media equation**

$$u_t = (u^n u_x)_x,$$

which occurs in many contexts. In (2.44) the flux is $Q = -D(u)u_x$, where the dispersion coefficient $D(u) = u$ depends on the concentration. We say the equation is degenerate because the dispersion coefficient has the property that $D(u) \to 0$ as $u \to 0$; it is not bounded away from zero. This degeneracy gives the equation distinctive properties that do not occur if $D(u) \geq \epsilon > 0$. We append to (2.44) the conditions

$$u(x,0) = 0, \quad x \neq 0, \tag{2.45}$$

$$\int_{\mathbb{R}} u(x,t)dx = 1, \quad t \neq 0. \tag{2.46}$$

These conditions represent the release of a unit amount of contaminant at the origin at $t = 0$ [in other symbols, $u(x, 0) = \delta(x)$, the delta distribution]. This problem admits a similarity solution of the form

$$u = t^{-1/3} y(s), \quad s = x/t^{1/3},$$

where

$$(yy')' + \frac{1}{3}(sy)' = 0.$$

Integrating once gives

$$y(y' + \frac{s}{3}) = 0.$$

We have set the constant of integration to equal zero because $y(\pm\infty) = 0$. It follows that

$$y = 0 \quad \text{or} \quad y = \frac{1}{6}(s_0^2 - s^2),$$

where s_0 is an arbitrary constant. Neither solution alone can satisfy the conditions (2.45)–(2.46) on $\mathbb{R}$, so we piece together these two solutions to obtain

$$u(x, t) = 0, \quad |s| > s_0; \quad u(x, t) = \frac{1}{6}(s_0^2 - s^2)t^{-1/3}, \quad |s| < s_0.$$

We choose s_0 so that

$$\int_{\mathbb{R}} y(s)\,dx = 1,$$

which gives $s_0 = 3^{2/3}/2^{1/3}$. Consequently,

$$u(x, t) = 0, \quad |x| > s_0 t^{1/3}; \quad u(x, t) = \frac{t^{-1/3}}{6}(s_0^2 - \frac{x^2}{t^{2/3}}), \quad |x| < s_0 t^{1/3}. \quad (2.47)$$

Figure 2.4 shows several time snapshots of the concentration. The solution is continuous, but u_x is not continuous. There are sharp "wavefronts" propagating along the spacetime loci $|x| = s_0 t^{1/3}$, much like one can observe in a hyperbolic problem. Thus, this is not a solution in the classical sense; it is an example of a *weak* solution.

Exercise 38 *Find solutions of the equation*

$$xu_t = u_{xx}, \quad x, t > 0$$

of the form $u = t^\gamma F(x/t^{1/3})$. Obtain the solution $u = t^{-2/3}\exp(-x^3/9t)$.

2.5.3   Blowup of Solutions

Another chacteristic phenomenon often occurring in nonlinear parabolic problems is blow up of solutions in finite time. In other words, the growth of the solution, measured in some manner, becomes infinite at a finite time. This reminds us of similar behavior for ordinary differential equations. For example,

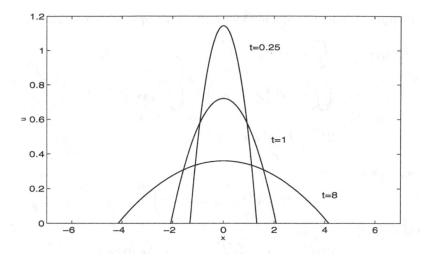

Figure 2.4: Time snapshots of the solution (2.47).

the initial value problem $y' = 1 + y^2$, $y(0) = 0$, has solution $y(t) = \tan t$, which blows up at $t = \pi/2$. To illustrate this type of behavior we consider the semilinear reaction–diffusion equation

$$
\begin{aligned}
u_t &= u_{xx} + u^3, \quad x \in (0, \pi), \quad t > 0, & (2.48) \\
u(0, t) &= u(\pi, t) = 0, \quad t > 0, & (2.49) \\
u(x, 0) &= u_0(x), \quad x \in (0, \pi). & (2.50)
\end{aligned}
$$

The initial datum u_0 is assumed to be continuous and nonnegative on $(0, \pi)$, and

$$
\int_0^\pi u_0(x) \sin x \, dx > 2.
$$

From the comparison principle we observe that the solution is nonnegative so long as it exists. We define

$$
s(t) = \int_0^\pi u(x, t) \sin x \, dx.
$$

Using integration by parts,

$$
\begin{aligned}
s'(t) &= \int_0^\pi u_t(x, t) \sin x \, dx \\
&= \int_0^\pi (u_{xx}(x, t) \sin x + u^3 \sin x) dx \\
&= -s(t) + \int_0^\pi u^3(x, t) \sin x \, dx.
\end{aligned}
$$

Now we apply Holder's inequality[2] with $p = 3$ and $q = 3/2$ to obtain

$$
\begin{aligned}
s(t) &= \int_0^\pi u \sin x \, dx = \int_0^\pi \sin^{2/3} x \, u \sin^{1/3} x \, dx \\
&\leq \left(\int_0^\pi (\sin^{2/3} x)^{3/2} \, dx \right)^{2/3} \left(\int_0^\pi (u \sin^{1/3} x)^3 \, dx \right)^{1/3} \\
&\leq 2^{2/3} \left(\int_0^\pi u^3 \sin x \, dx \right)^{1/3}.
\end{aligned}
$$

In other words,

$$
s(t) \leq 4 \left(\int_0^\pi u^3 \sin x \, dx \right)^{1/3},
$$

and therefore

$$
s'(t) \geq -s(t) + \frac{s(t)^3}{4}, \quad t > 0, \quad s(0) > 2.
$$

This inequality will imply that $s(t) \to \infty$ at a finite time t. To show this is the case let $v = 1/s^2$. Then the differential inequality becomes

$$
v'(t) \leq 2v(t) - \frac{1}{2}.
$$

Multiplying by $\exp(-2t)$ and integrating gives

$$
v(t) \leq \frac{1}{4}(1 - e^{-2t}) + v(0)e^{2t}.
$$

Consequently,

$$
s(t) \geq \left(e^{2t}(\frac{1}{s(0)^2} - \frac{1}{4}) + \frac{1}{2} \right)^{-2}.
$$

Because $s(0) > 2$, the right-side of this inequality goes to infinity at a finite value of t. Therefore the solution blows up in finite time.

Reaction-diffusion equations can have this type of behavior where solutions only exist locally, for finite times. The situation would not improve if we added a linear advection term. This type of phenomenon is characteristic for some equations containing reaction terms. In (2.48)–(2.50) the presence of diffusion, or dispersion, does not cause decay; reaction wins the competition and the solution blows up. Blow up for reaction diffusion equations is discussed thoroughly in Samarski *et al* (1995).

2.5.4   Stability of the Zero Solution

Another method to aid in understanding the behavior of nonlinear evolution equations is to inquire about the linearized stability of steady-state solutions. This study is really about the permanence of steady solutions when they are

---

[2]If $1/p + 1/q = 1$, then $\int |fg| dx \leq (\int |f|^p dx)^{1/p} (\int |g|^q dx)^{1/q}$.

subjected to small perturbations. Let us consider the equilibrium model on a bounded domain:

$$(u + \beta f(u))_t = \alpha u_{xx} - u_x, \quad 0 < x < 1, \, t > 0, \qquad (2.51)$$
$$u(0,t) = u(1,t) = 0, \quad t > 0, \qquad (2.52)$$

where f satisfies the conditions $f(0) = 0, f \in C^1[0,\infty) \cup C^2(0,\infty)$, with $f > 0$, $f' > 0$, $f'' < 0$. It is clear that $u \equiv 0$ is a steady-state solution to the problem. If $w = w(x,t)$ represents a small perturbation to the zero solution, then $f(w) = f(0) + f'(0)w + \frac{1}{2}f''(\tilde{w})w^2$, where $0 < \tilde{w} < w$, and the perturbation w satisfies the linearized problem

$$(w + \beta f'(0)w)_t = \alpha w_{xx} - w_x, \quad 0 < x < 1, \, t > 0, \qquad (2.53)$$
$$w(0,t) = w(1,t) = 0, \quad t > 0. \qquad (2.54)$$

The linearized equation (2.53) can be written

$$w_t = dw_{xx} - aw_x, \quad 0 < x < 1, \, t > 0,$$

where $d = \alpha/(1 + f'(0))$ and $a = 1/(1 + f'(0))$. To analyze this equation we use an **energy method**. First multiply the equation by w and then integrate over $0 < x < 1$ to get

$$\int_0^1 ww_t dx = \int_0^1 dww_{xx} dx - \int_0^1 aww_x dx.$$

Now observe that $2ww_t = (w^2)_t$, $2ww_x = (w^2)_x$, and integrate the first term on the right-side by parts. We obtain

$$\frac{d}{dt}\int_0^1 w^2 dx = 2dww_x \big|_0^1 - \int_0^1 w_x^2 dx - w^2 \big|_0^1 .$$

But the boundary conditions (2.54) force the two boundary terms to vanish and we obtain

$$\frac{d}{dt}\int_0^1 w^2 dx = -2d \int_0^1 w_x^2 dx.$$

Therefore

$$\frac{d}{dt}\|w(\cdot,t)\|^2 \le 0,$$

where

$$\|w(\cdot,t)\| = \left(\int_0^1 w^2 dx\right)^{1/2}$$

is the $L^2[0,1]$ or energy norm. Therefore, the energy norm of small perturbations remain bounded for all $t > 0$.

The previous calculation shows that small perturbations stay under control when a linearized analysis is performed. We can sometimes analyze the nonlinear equation in the same manner, using an energy method. Note that the nonlinear equation can be written

$$g(u)_t = \alpha u_{xx} - u_x,$$

where $g(u) = u + \beta f(u) > 0$. Multiplying by $g(u)$ and integrating from over $0 < x < 1$ yields

$$
\begin{aligned}
\frac{d}{dt}\|g(u(\cdot,t))\|^2 &= 2\alpha \int_0^1 g(u)u_{xx}dx - 2\int_0^1 g(u)u_x dx \\
&= 2\alpha g(u)u_x \mid_0^1 -2\alpha \int_0^1 g'(u)u_x^2 dx - 2\int_0^1 g(u)u_x dx \\
&= -2\alpha \int_0^1 g'(u)u_x^2 dx - 2\int_0^1 g(u)u_x dx.
\end{aligned}
$$

Because $g'(u) = 1 + f'(u) > 0$, the first term on the right is nonpositive. To estimate the second term let $G(u)$ be the antiderivative of g, i.e., $G(u) = \int_0^u g(y)dy$. Then

$$\int_0^1 g(u)u_x dx = \int_0^1 G(u)_x dx = G(u) \mid_0^1 = 0.$$

Therefore

$$\frac{d}{dt}\|g(u(\cdot,t))\|^2 = -2\alpha \int_0^1 g'(u)u_x^2 dx \le 0,$$

and so the norm of $g(u)$ stays under control. Thus $\|u(\cdot,t)\|$ remains bounded and the energy stays under control.

Arguments like those given above are predicated on the assumption that solutions exist and are called *a priori* estimates.

2.6   The Reaction–Advection Equation

2.6.1   Semilinear Equations

For completeness, we now examine some properties of the preceding equations when dispersion is absent, i.e., $\alpha = 0$, and when the reaction term is given by $\Phi = \Phi(u)$. In deep bed filtration theory (Chapter 3), we shall observe that the dispersion term is neglected in some models. With neglect of dispersion, equation (2.3) becomes the **reaction–advection equation**

$$u_t + vu_x = \Phi(u), \tag{2.55}$$

which is, in general, a semilinear hyperbolic equation. Therefore, this equation is not parabolic at all and we do not have some of the properties expected

in parabolic equations.   To solve reaction-advection equations we introduce a
moving coordinate system, or characteristic coordinates, defined by

$$\xi = x - vt, \quad \tau = t.$$

Then the equation becomes

$$U_\tau = \Phi(U),$$

where $U(\xi, \tau) = u(\xi + v\tau, \tau)$.   (Easily, by the chain rule for derivatives, the
advection operator $\partial_t + v\partial_x$ simplifes to just ∂_τ in characteristic coordinates.)
Therefore,

$$\int^U \frac{dw}{\Phi(w)} = \tau + \psi(\xi),$$

where ψ is an arbitrary function. Thus, the equation

$$\int^u \frac{dw}{\Phi(w)} = t + \psi(x - vt)$$

defines, implicitly, the general solution of (2.55).   The arbitrary function is
determined specifically by initial or boundary data. The following exercise and
example give two illustrations: a Cauchy problem with initial datum given on
the entire real line and a Cauchy–Dirichlet problem where initial datum is given
on a half line and a Dirichlet condition is prescribed at the boundary.

Exercise 39 *Consider the Cauchy problem for the reaction–advection equation
with Langmuir kinetics:*

$$u_t + vu_x \;=\; -\frac{k_1 u}{k_2 + u}, \quad x \in \mathbb{R},\ t > 0, \tag{2.56}$$

$$u(x, 0) \;=\; u_0(x), \quad x \in \mathbb{R}, \tag{2.57}$$

where $u_0(x) \geq 0$. Show that the solution is given implicitly by

$$-t = \int_{u_0}^{u} \left(\frac{k_2}{k_1 w} + \frac{1}{k_1} \right) dw,$$

or, after simplification,

$$u - k_1 t - u_0 = k_2 \ln(u/u_0).$$

*Using graphical techniques show that for each x and $t > 0$ there is a unique
value $u = u(x, t) < u_0(x)$. For $t > 0$ the single root at $t = 0$ bifurcates into two
roots, the smaller of which is the solution. As $t \to 0+$ it is clear that $u \to 0$.
Therefore show that the Cauchy problem (2.56)–(2.57) has a global solution.*

Example 40 *Consider the Cauchy–Dirichlet problem for the reaction-advection
equation with Freundlich-type kinetics:*

$$u_t + vu_x \;=\; -\sqrt{u}, \quad x > 0,\ t > 0, \tag{2.58}$$

$$u(x, 0) \;=\; u_0(x), \quad x > 0, \tag{2.59}$$

$$u(0, t) \;=\; g(t), \quad t > 0, \tag{2.60}$$

where $u_0(x) \geq 0$ and $g(t) \geq 0$. In characteristic coordinates the equation becomes

$$U_\tau = -\sqrt{U}.$$

Integration yields the general solution

$$\sqrt{U} = -\frac{\tau}{2} + \psi(\xi)$$

or

$$\sqrt{u} = -\frac{t}{2} + \psi(x - vt) \tag{2.61}$$

where ψ is an arbitrary function. Because signals in this system travel at speed v, we treat the regions $x > vt$ and $x < vt$ separately. The region $x > vt$, which is ahead of the leading signal $x = vt$, is influenced by the initial data (2.59) along $t = 0$. Thus, applying the initial condition to the general formula for u gives

$$\sqrt{u_0(x)} = \psi(x),$$

thereby determining the arbitrary function for this region. Therefore

$$\sqrt{u} = -\frac{t}{2} + \sqrt{u_0(x - vt)}, \quad x > vt.$$

For the region $0 < x < vt$ we apply the boundary condition to the general solution to get

$$\sqrt{g(t)} = -\frac{t}{2} + \psi(-vt).$$

Thus, replacing $-vt$ by $t - x/v$,

$$\psi(x - vt) = \frac{t - x/v}{2} + \sqrt{g(t - x/v)}.$$

Therefore,

$$\sqrt{u} = -\frac{x}{2v} + \sqrt{g(t - x/v)}, \quad 0 < x < vt.$$

We observe that these solution formulas hold only for those times for which the right-hand sides are nonnegative. For times greater than the loci where the right-hand sides vanish, we set $u(x, t) \equiv 0$.

2.6.2   Quasilinear Equations

In the absence of dispersion, the equilibrium model (2.22)–(2.23) becomes

$$(1 + \beta f'(u))u_t + u_x = 0. \tag{2.62}$$

Unlike a semilinear equation where the nonlinearity appears in the reaction, or source term, this equation is quasilinear and the nonlinearity occurs in the differential operator. We can anticipate the development of singularities (shocks) as a solution propagates in time.

We rewrite (2.62) as

$$u_t + c(u)u_x = 0,$$

where

$$c(u) = \frac{1}{1 + \beta f'(u)}.$$

Observe, from the assumptions on the isotherm f, that

$$c(u) > 0 \quad c'(u) > 0.$$

Thus we have a standard kinematic wave equation. The analysis of such equations is straightforward and can be found in many texts [e.g., see Logan (1997c, 1994), Whitham (1974), or Smoller (1994)].

To illustrate the analysis, let us consider the Cauchy problem, or the pure initial value problem on $\mathbb{R}$. We impose the initial concentration

$$u(x, 0) = u_0(x), \quad x \in \mathbb{R}.$$

We define the characteristic curves as solutions of the equation

$$\frac{dx}{dt} = c(u(x, t)).$$

On these curves the PDE becomes

$$\frac{du}{dt} = 0 \quad \text{or} \quad u = \text{const.}$$

From the calculation

$$\frac{d^2x}{dt^2} = \frac{d}{dt}c(u) = c'(u)\frac{du}{dt} = 0,$$

it follows that the characteristic curves are straight lines. If (x, t) is an arbitrary point in spacetime, then the characteristic line connecting (x, t) to a point $(\xi, 0)$ on the x axis has speed $c(u_0(\xi))$ and is given by

$$x - \xi = c(u_0(\xi))t. \tag{2.63}$$

By the constancy of u on the characteristic, the solution u at (x, t) is given by

$$u(x, t) = u_0(\xi). \tag{2.64}$$

The two equations (2.63)–(2.64) define the solution, if it exists; the solution is (2.64) where $\xi = \xi(x, t)$ is defined implicitly by (2.63). To determine the validity of the solution, let us calculate the partials u_x and u_t. We have

$$u_x = u_0'(\xi)\xi_x = \frac{u_0'(\xi)}{1 + tc'(u_0(\xi))u_0'(\xi)},$$

where ξ_x was computed from (2.63). A similar calculation shows

$$u_t = u_0'(\xi)\xi_t = -\frac{c(u_0(\xi))u_0'(\xi)}{1 + tc'(u_0(\xi))u_0'(\xi)}.$$

It is easy to verify that $u_t + c(u)u_x = 0$, provided the denominator in the expression for the derivatives never vanishes, i.e.,

$$1 + tc'(u_0(\xi))u_0'(\xi) \neq 0.$$

Because $c'(u) > 0$, if the initial concentration $u_0(x)$ is nondecreasing, then the denominator is always positive and the solution to the Cauchy problem exists for all time. On the other hand, if there is a value of x_0 where $u_0'(x_0) < 0$, then there will be two characteristic lines that cross, contradicting the constancy of u on the characteristics. In this case the solution will blow up and a "gradient catastrophe" will occur in finite time; at this time the classical solution must terminate. At this blowup time a shock will form and a weak solution will be propagated.

2.7   Examples

2.7.1   Advection–Dispersion Equation

In Chapter 1 we discussed several properties of the diffusion, or dispersion, equation. In this section we solve some sample problems associated with the advection-dispersion equation

$$C_t = DC_{xx} - vC_x \tag{2.65}$$

and some of its variants. As we have noted, the equation can be put into dimensionless form

$$u_t = \alpha u_{xx} - u_x, \tag{2.66}$$

where α is the inverse of the Peclet number, i.e., Pe$= \alpha^{-1} = Lv/D$, where L is a length for the problem.

As in the case of the dispersion equation we can inquire about plane wave solutions of (2.66) of the form $u = e^{i(kx-\omega t)}$. Substituting into the equation (2.66) we find that the wave number k and frequency ω are related by the dispersion relation $\omega = k - \alpha k^2 i$. Therefore, plane wave solutions have the form

$$u(x,t) = e^{-\alpha k^2 t}e^{ik(x-t)},$$

which are oscillatory, decaying, waves moving at the unit advection speed.

The fundamental solution of the advection-dispersion equation (2.66) is the solution of the Cauchy problem

$$
\begin{aligned}
u_t &= \alpha u_{xx} - u_x, \quad x \in \mathbb{R},\ t > 0 \tag{2.67}\\
u(x,0) &= \delta(x), \quad x \in \mathbb{R}, \tag{2.68}
\end{aligned}
$$

where δ is the delta distribution. Hence, it is the response to a unit, point source at $x = 0$ applied at the time $t = 0$. The simplest way to solve (2.67)–(2.68) is to transform to characterisitic, moving coordinates $\xi = x - t$, $\tau = t$. Then the problem becomes

$$u_\tau = \alpha u_{\xi\xi}, \quad \xi \in \mathbb{R}, t > 0, \tag{2.69}$$
$$u(\xi, 0) = \delta(\xi), \quad \xi \in \mathbb{R}, \tag{2.70}$$

which is the Cauchy problem for the diffusion equation. By the results in Section 1.1 we have

$$u(\xi, \tau) = g(\xi, \tau),$$

where g is the fundamental solution of the diffusion equation. Consequently, the fundamental solution of the advection-dispersion equation is

$$u(x, t) = g(x - t, t) = \frac{1}{\sqrt{4\pi\alpha t}} e^{-(x-t)^2/4\alpha t}.$$

By superimposing the responses caused by a distributed initial source ϕ, we can obtain the solution to the general Cauchy problem

$$u_t = \alpha u_{xx} - u_x, \quad x \in \mathbb{R}, t > 0, \tag{2.71}$$
$$u(x, 0) = \phi(x), \quad x \in \mathbb{R}, \tag{2.72}$$

as

$$u(x, t) = \int_{-\infty}^{\infty} \frac{1}{\sqrt{4\pi\alpha t}} e^{-(x-y-t)^2/4\alpha t} \phi(y) dy. \tag{2.73}$$

Exercise 41 *In the case where the initial condition is Gaussian, i.e., $\phi(x) = e^{-x^2/a}$, show that the integral in (2.73) can be calculated in analytic form to obtain*

$$u(x, t) = \frac{\sqrt{a}}{\sqrt{a + 4\alpha t}} e^{-(x-t)^2/(a+4\alpha t)}.$$

Conclude that, as time increases, the Gaussian concentration profile decays, advects to the right with speed one, and spreads outward.

Exercise 42 *If the initial condition is a step function $\phi(x) = 1 - H(x)$, where H is the Heaviside function, show that the solution to (2.71)–(2.72) is*

$$u(x, t) = \frac{1}{2} \left(1 + \mathrm{erf} \left(\frac{t - x}{\sqrt{4\alpha t}} \right) \right), \quad x < t,$$

and

$$u(x, t) = \mathrm{erf}\, c \left(\frac{x - t}{\sqrt{4\alpha t}} \right), \quad x > t,$$

where erfc $= 1 - $ erf.

2.7.2   Boundary Conditions

Problems with boundaries require boundary data. In addition to the usual Dirichlet condition, where the concentration is specified on a boundary, there are other important boundary conditions that arise from the fact that the flux has two parts, an advective part caused by the bulk flow and a dispersive part caused by diffusion and mechanical dispersion. Every problem must be analyzed carefully to determine which condition best models the physical situation.

An important boundary condition at an inlet boundary arises naturally for the advection–dispersion equation (2.65). Assume that the problem is defined on a semi-infinite domain $x > 0$ with inlet boundary at $x = 0$. The flux is given by $Q \equiv -DC_x + vC$, and if we require that the flux be continuous across the boundary, we have

$$-DC_x(0^+, t) + vC(0^+, t) = -DC_x(0^-, t) + vC(0^-, t),$$

where 0^+ and 0^- denote the right and left limits, respectively. If the region to the left of the boundary is a reservoir (e.g., a well or a lake) where the chemical has concentration $C_0(t)$ and is perfectly stirred, then there are no gradients and we have

$$C_x(0^-, t) = 0, \quad C(0^-, t) = C_0(t).$$

Thus, we obtain the **Fourier boundary condition**

$$-DC_x(0, t) + vC(0, t) = vC_0(t), \quad t > 0.$$

If $x = l$ is an outflow boundary, we can make a similar argument as in the last example and equate the flux on both sides of the boundary, or

$$-DC_x(l^+, t) + vC(l^+, t) = -DC_x(l^-, t) + vC(l^-, t).$$

Assuming the region to the right of $x = l$ is a perfectly stirred reservoir and that its concentration is the same as the concentration exiting the domain, then we get the condition

$$C_x(l, t) = 0, \quad t > 0.$$

Observe that this is *not* a no-flux condition; it must be remembered that the flux involves an advection term that is not zero. This boundary condition is called a zero-gradient condition.

Exercise 43 *The advection–dispersion equation (2.65) on a bounded domain can be solved by the eigenfunction expansion method (see Chapter 1). Consider the problem with Dirichlet boundary conditions:*

$$
\begin{aligned}
C_t &= DC_{xx} - vC_x, \quad 0 < x < l, \, t > 0, \\
C(0, t) &= C(l, t) = 0, \quad t > 0, \\
C(x, 0) &= f(x), \quad 0 < x < l.
\end{aligned}
$$

Show that the associated Sturm–Liouville problem is

$$Dy'' - vy' = \lambda y, \quad 0 < x < l; \quad y(0) = y(l) = 0,$$

and that the eigenvalues and eigenfunctions are given by

$$\lambda_n = -\frac{v^2 l^2 + 4n^2 \pi^2 D^2}{4Dl^2}, \quad y_n(x) = e^{vx/2D} \sin \frac{n\pi x}{l}, \quad n = 1, 2, \ldots$$

Thus, obtain the solution

$$u(x,t) = \sum_{n=1}^{\infty} a_n e^{\lambda_n t} e^{vx/2D} \sin \frac{n\pi x}{l},$$

where

$$a_n = \frac{(f, y_n)}{\|y_n\|^2}.$$

In Chapter 1 we introduced a pure boundary value problem associated with the diffusion equation. Such a problem models the flow in a half-space when boundary conditions are imposed for a long time. We can proceed in the same manner for more general equations. Consider the two boundary value problems

$$c_t = Dc_{xx} - vc_x - \lambda c, \quad x > 0, \ t \in \mathbb{R}, \quad (2.74)$$
$$-Dc_x(0,t) + vc(0,t) = vf(t), \quad t \in \mathbb{R}, \quad (2.75)$$

and

$$u_t = Du_{xx} - vu_x - \lambda u, \quad x > 0, \ t \in \mathbb{R}, \quad (2.76)$$
$$u(x,0) = f(t), \quad t \in \mathbb{R}, \quad (2.77)$$

where the first has a Fourier boundary condition and the second has a Dirichlet condition. It is straightforward to verify that bounded solutions to the two problems are connected by the relation

$$c(x,t) = \frac{v}{D} e^{vx/D} \int_x^\infty u(y,t) e^{-vy/D} dy,$$

which effectively allows the reduction of a problem with a Fourier condition to one with a Dirichlet condition. The connection is also defined by the differential relation

$$vu(x,t) = -Dc_x(x,t) + vc(x,t).$$

It is well-known [e.g., see Guenther and Lee (1996)] that the solution to (2.76)–(2.77) is

$$u(x,t) = \frac{2}{\sqrt{\pi}} e^{vx/2D} \int_0^\infty e^{-y^2 - \frac{(\lambda + v^2/4D)x^2}{4Dy^2}} f\left(t - \frac{x^2}{4Dy^2}\right) dy.$$

Observe that if f is a periodic function, then so is u. Further results on periodic boundary conditions and other references can be found in Logan and Zlotnik (1995).

Exercise 44 *Verify the details in the last few paragraphs.*

2.7.3   A Perturbation Problem

Perturbation methods provide a powerful technique to obtain approximate solutions to difficult problems when a large or small parameter is present. In the next few paragraphs we use perturbation methods to analyze a simple problem that can be solved analytically by other methods. Consider the problem

$$C_t = DC_{xx} - vC_x - \lambda C, \quad x > 0, \ t > 0, \tag{2.78}$$
$$C(0,t) = C_b(t), \quad t > 0, \tag{2.79}$$
$$C(x,0) = 0, \quad x > 0, \tag{2.80}$$

that models the advection, dispersion, and decay of a chemical tracer on the semi-infinite domain $x > 0$, with a given Dirichlet boundary condition. Furthermore, assume that the dispersion constant is small in some sense (which we clarify later). We have already observed that this problem can be transformed into one involving the diffusion equation by letting

$$C(x,t) = u(x,t)e^{vx/2D-(\lambda+v^2/4D)}.$$

Then u satisfies

$$u_t = Du_{xx}, \quad x > 0, \ t > 0,$$
$$u(0,t) = h(t) \equiv C_b(t)e^{(\lambda+v^2/4D)t}, \quad t > 0,$$
$$u(x,0) = 0, \quad x > 0.$$

Using Laplace transforms, the solution to this problem is

$$u(x,t) = -2D \int_0^t g_x(x,t-\tau)h(\tau)d\tau,$$

where $g(x,t)$ is the fundamental solution. Therefore we have obtained the exact solution to the problem, but it is somewhat obscured because of the complicated integral formula. We can often get a better idea of the behavior of the solution using a singular perturbation method. Such a strategy is common; asymptotic methods are often preferred over difficult integral representations (which, by the way, usually require asymptotic approximations anyway). The reader unfamiliar with singular perturbation methods can consult Kevorkian and Cole (1981), Lin and Segel (1989), or Logan (1997c).

First we recast the problem into dimensionless form. Scaling t by λ^{-1}, x by v/λ, and C by the maximum value of $C_b(t)$, we obtain the model

$$c_t = \epsilon c_{xx} - c_x - c, \quad x > 0, \ t > 0, \tag{2.81}$$
$$c(0,t) = c_b(t), \quad t > 0, \tag{2.82}$$
$$c(x,0) = 0, \quad x > 0, \tag{2.83}$$

where

$$\epsilon = \frac{\lambda D}{v^2}.$$

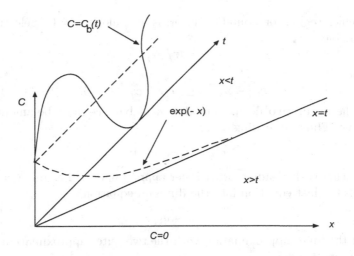

Figure 2.5: Spacetime diagram.

The assumption that the dispersion constant is small means precisely that $\epsilon <<$ 1, or the constant D is small compared to v^2/λ.

When we set $\epsilon = 0$, we obtain the **outer problem**

$$c_{0t} = -c_{0x} - c_0,$$

which is a hyperbolic equation. The general solution to this simple advection equations is (see Section 2.4)

$$c_0(x,t) = G(x-t)e^{-t},$$

where G is an arbitrary function. In the domain $x > t$, i.e., ahead of the leading signal from the boundary, we clearly have

$$c_0(x,t) = 0, \quad x > t.$$

Behind the wave, i.e., for $0 < x < t$ we apply the boundary condition to determine $G(t) = c_b(-t)e^{-t}$. Therefore, behind the wave we have

$$c_0(x,t) = c_b(t-x)e^{-x}, \quad 0 < x < t.$$

Thus, the outer solution is defined in two pieces. Along the leading edge $x = t$ we expect exponential decay because $c_0(t^-,t) = c_b(0)e^{-x} = e^{-x}$. But we note that the two solutions do not match along the line $x = t$. It is here, in a neighborhood of $x = t$, that we require an "inner" approximation that will tie together the two pieces of the outer approximation. Figure 2.5 shows depicts the situation geometrically.

To find the inner approximation we change to characteristic coordinates: $\eta = x$, $\tau = t - x$. In these variables the PDE becomes

$$c_{\eta\eta} - 2\epsilon c_{\eta\tau} + \epsilon c_{\tau\tau} - c_\eta - c = 0,$$

and the inner region, or boundary layer, is now along $\tau = 0$. Selecting a new scaled variable

$$\xi = \tau/\sqrt{\epsilon},$$

we obtain

$$\epsilon c_{\eta\eta} - 2\sqrt{\epsilon}c_{\eta\xi} + c_{\xi\xi} - c_\eta - c = 0,$$

which is the **inner problem**. The dominant balance must be among the last three terms. Thus, to leading order,

$$c^i_{\xi\xi} - c^i_\eta - c^i = 0,$$

where c^i denotes the leading-order inner approximation. Letting $u = c^i e^\eta$ then transforms the last equation into the diffusion equation

$$u_\eta = u_{\xi\xi}.$$

Matching the inner approximation with the two outer approximation gives the boundary conditions

$$c^i \to 0 \quad \text{as } \xi \to -\infty$$

and

$$c^i \to e^{-\eta} \quad \text{as } \xi \to +\infty.$$

Thus, $u \to 0$ as $\xi \to -\infty$ and $u \to 1$ as $\xi \to +\infty$. Therefore, the solution to the u problem is

$$u(\xi, \eta) = \frac{1}{2}\left(1 + \text{erf}\left(\frac{\xi}{\sqrt{4\eta}}\right)\right).$$

Hence

$$c^i(\tau, \eta) = \frac{1}{2}(1 + \text{erf}\left(\frac{\tau/\sqrt{\epsilon}}{\sqrt{4\eta}}\right)e^{-\eta}.$$

Returning to the original coordinates, we have inner approximation

$$c^i(x, t) = \frac{1}{2}\left(1 + \text{erf}\left(\frac{t-x}{\sqrt{4\epsilon x}}\right)\right)e^{-x}.$$

This is the approximation that joins the two pieces of the outer approximation.

Finally, we can form a uniform approximation by adding the inner and outer approximations and then subtracting their common limit. We obtain

$$c(x, t) = \frac{1}{2}\left(1 + \text{erf}\left(\frac{t-x}{\sqrt{4\epsilon x}}\right)\right)e^{-x}, \quad x > t,$$

$$c(x, t) = c_b(t-x)e^{-x} + \frac{1}{2}\left(1 + \text{erf}\left(\frac{t-x}{\sqrt{4\epsilon x}}\right)\right)e^{-x} - e^{-x}, \quad 0 < x < t.$$

Exercise 45 *Find bounded solutions to the following model with spatially dependent dispersion:*

$$c_t = ((a + vx)c_x)_x - vc_x - \lambda c, \quad x > 0, \quad t \in \mathbb{R},$$
$$c(0, t) = \sin\omega t, \quad t \in \mathbb{R},$$

and describe how the amplitude and phase depend on ω. *Hint: assume that* $c = \phi(x)e^{i\omega t}$ *and then change the independent variable to* $\xi = a + vx$.

2.7.4 Radial Dispersion

The one-dimensional advection–dispersion equation in the semi-infinite domain $x > 0$ is

$$u_t = \alpha u_{xx} - u_x, \quad x > 0, \ t > 0.$$

With initial and boundary conditions given by

$$u(x,0) = 0, \ x > 0; \quad u(0,t) = u_0 = \text{const.},$$

it is well-known [e.g., see Sun (1995)] that the solution is given by

$$u(x,t) = \frac{u_0}{2} \left(\text{erf}\,c \left(\frac{x-t}{\sqrt{4\alpha t}} \right) + e^{x/\alpha} \ \text{erf}\,c \left(\frac{x+t}{\sqrt{4\alpha t}} \right) \right).$$

Here, erf c is the complementary error function defined by erf $c\,(z) = 1 - \text{erf}\,(z)$, where $\text{erf}(z)$ is the error function. For each fixed x, the second term in the solution becomes negligible quickly and so the solution is often approximated by first term.

The solution to the one-dimensional advection–dispersion equation with decay, subject to the same initial and boundary data, can be found in, for example, de Marsily (1986). Many other solutions are given Sun (1995). Most of these analytic solutions are found using transform methods. A compendium is given in van Genuchten and Alves (1982).

In the next few paragraphs we examine a simple advection-dispersion equation in radial geometry:

$$c_t = \frac{\alpha a}{r} c_{rr} - \frac{a}{r} c_r, \quad r > r_0, \ t > 0, \tag{2.84}$$

$$c(r,0) = 0, \ r > r_0, \quad c(r_0, t) = c_0, \ t > 0. \tag{2.85}$$

We assume that solutions remain bounded as $r \to 0$. Rescaling the problem via

$$\bar{r} = \frac{r}{\alpha}, \ \bar{t} = \frac{t}{\alpha^2/a}, \ u = \frac{c}{c_0}$$

gives, upon dropping the overbars, the dimensionless model

$$u_t = \frac{1}{r} u_{rr} - \frac{1}{r} u_r, \quad r > R, \ t > 0, \tag{2.86}$$

$$u(r,0) = 0, \ r > R, \quad u(R,t) = 1, \ t > 0, \tag{2.87}$$

where $R \equiv r_0/\alpha$.

Now let $U(r,s)$ denote the Laplace transform of $u(r,t)$. Then, taking Laplace transforms of the PDE and boundary condition gives

$$U'' - U' - srU = 0, \quad U(R,s) = \frac{1}{s},$$

where prime denotes the r derivative. The first derivative term can be eliminated by making a transformation of the dependent variable to

$$W = U e^{-r/2}.$$

Then the differential equation becomes

$$W'' - \left(\frac{1}{4} + sr\right) W = 0.$$

Changing the independent variable to

$$z = (\frac{1}{4} + sr)/s^{2/3}$$

transforms the equation into **Airy's equation**

$$\frac{d^2 W}{dz^2} - zW = 0.$$

Here we have used the same symbol for the dependent variable W. The general solution to Airy's equation is [see Abramowitz and Stegun (1962)]

$$W(z, s) = c_1(s) \text{ Ai}(z) + c_2(s) \text{ Bi}(z).$$

To keep solutions bounded we must set $c_2(s) = 0$. Hence,

$$W(r, s) = c_1(s) \text{ Ai}\left(\frac{0.25 + sr}{s^{2/3}}\right),$$

or

$$U(r, s) = c_1(s) e^{r/2} \text{ Ai}\left(\frac{0.25 + sr}{s^{2/3}}\right).$$

Finally we apply the boundary condition $U(R, s) = 1/s$ and we obtain the solution to the problem (2.86)–(2.87) in the transform domain:

$$U(r, s) = \frac{1}{s} e^{(r-R)/2} \text{ Ai}(\frac{0.25 + sr}{s^{2/3}})/ \text{ Ai}(\frac{0.25 + sR}{s^{2/3}}).$$

Although the inversion appears formidable (and it is in the analytic sense), it is relatively easy if carried out numerically. There are several numerical algorithms available to invert Laplace transforms, and two of them are discussed briefly in Appendix C. Here we use the **Stehfest algorithm**. The Maple V (version 5) worksheet, presented in Appendix C, can produce the plots shown in figure 2.6, which are two profiles of the concentration $u(r, t)$. The scaled well radius is $R = 1$.

Exercise 46 *Exercise 47 Consider the model advection–adsorption–dispersion-decay model*

$$u_t = au_{xx} - u_x + s - c,$$
$$s_t = u - s,$$

on the domain $x, t > 0$ with $u(x, 0) = s(x, 0) = 0$ and $u(0, t) = u_0$. Take Laplace transforms and show that the bounded solution in the transform domain is

$$U(x, p) = \frac{u_0}{p} e^{mx},$$

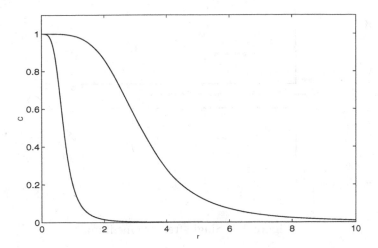

Figure 2.6: Two radial concentration profiles showing how a solute disperses and advects from a well.

where

$$m = \frac{1}{2a} - \frac{1}{\sqrt{a}}\sqrt{\frac{1}{4a} + \frac{p(p+2)}{p+1}}.$$

Obtain an approximation for $t \gg 1$ (large t) and for $t \ll 1$ (small t) by considering $p \ll 1$ and $p \gg 1$, respectively. Show that

$$U(x,p) \sim \frac{u_0}{p} e^{x/2a} e^{-\frac{x}{\sqrt{a}}\sqrt{\frac{1}{4a} + ks}},$$

where $k = 1$ for $p \gg 1$ and $k = 2$ for $p \ll 1$. Thus, show

$$u(x,t) \sim \frac{u_0}{2\pi i} e^{x/2a} \int_B e^{-\frac{x}{\sqrt{a}}\sqrt{\frac{1}{4a} + ks}} e^{pt} dp,$$

where $k = 1$ for $t \ll 1$ and $k = 2$ for $t \gg 1$. Here, B is a Bromwich path. Finally, use a table of Laplace transforms to invert the last expression, obtaining the approximation

$$u(x,t) \sim \frac{u_0}{2}\left\{ \operatorname{erf} c\left[\frac{1}{2\sqrt{a}}\left(x\sqrt{\frac{k}{t}} - \sqrt{\frac{t}{k}}\right)\right]\right.$$
$$\left. + e^{x/a} \operatorname{erf} c\left[\frac{1}{2\sqrt{a}}\left(x\sqrt{\frac{k}{t}} + \sqrt{\frac{t}{k}}\right)\right]\right\}$$

where $k = 1$ for small t and $k = 2$ for large t.

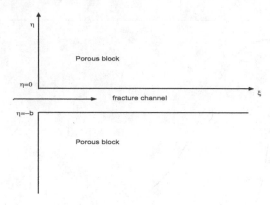

Figure 2.7: Single-fracture medium.

2.7.5   Fractured Media

Fractures in porous media are rapidly conducting pathways that channel the water through the domain. The presence of fractures in the porous fabric can have a significant influence on groundwater transport. For example, it is reported in the literature that fracture networks can significantly retard the transport of migrating radionuclides through the medium. Unfortunately, the structure of fracture systems can be quite complex and difficult to model. Therefore, many of the mathematical models developed in the literature apply only to idealized conditions and fracture geometries. These studies using simplified fissure structures are often considered worst-case scenarios in that, for example, a porous region with a single fracture will retain more of a decaying contaminant than will a complicated fracture network and will therefore lead to an overestimate of the retardation capacity of the porous blocks.

The simplified geometry we consider (figure 2.7) is the same as that studied by Tang *et al.* (1981) and others, namely a single fracture channel of width b in an infinite porous domain [also see Sun (1995), p. 69]. For the final form of the model, in the fracture we consider advection, decay, linear adsorption to the fracture surface, and loss to the fabric; in the porous blocks, or fabric, that bound the channel we impose unidirectional dispersion, perpendicular to the fracture channel, linear adsorption, and decay. This mathematical model leads to a single, degenerate, parabolic problem in a quarter-space with evolutionary, oblique boundary data and source terms.

To fix the notation, let $m(\xi, \eta, \tau)$ denote the concentration of a chemical contaminant (mass per unit volume of water) in the porous block $\xi, \eta > 0$. Here, τ is time. For the present, the contaminant is assumed to diffuse in both the ξ and η directions, and it experiences first-order decay and linear adsorption throughout the region. In the standard way, mass balance of the contaminant gives the governing equation

$$\omega m_\tau = D_1 m_{\xi\xi} + D_2 m_{\eta\eta} - \omega \Lambda m - \rho_b s_\tau,$$

where ω is the constant porosity, D_1 and D_2 are the molecular diffusion constants in the ξ and η directions, Λ is the decay constant, $\rho_b = \rho(1 - \omega)$ is the bulk density of the solid porous matrix, and $s = s(\xi, \eta, \tau)$ is the sorbed concentration given in mass per unit mass of solid fabric. We assume linear adsorption, that is, $s = K_d m$, where K_d is the distribution constant. We may combine the conditions above to write the mass balance equation in the porous domain as

$$R'm_\tau = D'm_{\xi\xi} + D''m_{\eta\eta} - \Lambda m, \quad \xi, \eta > 0, \quad \tau > 0, \tag{2.88}$$

where $R' = 1 + \rho_b K_d/\omega$ is the retardation constant, $D' = D_1/\omega$, $D'' = D_2/\omega$. It is clearly sufficient, by our symmetry assumption, to formulate and solve the problem in the quarter–space $\xi > 0, \eta > 0$.

In the fracture channel $-b < \eta < 0$, $\xi > 0$, the concentration, measured in mass per unit volume of water, is given by $c(\xi, \tau)$. Thus, we are assuming no variation in the concentration across the fracture. To derive the governing equation in the fracture we take a small section of the fracture channel of length $\Delta\xi$, width b, and unit thickness. Mass balance applied to this section yields

$$\frac{\partial}{\partial t}(cb\Delta\xi) = b(\phi(\xi, \tau) - \phi(\xi + \Delta\xi, \tau)) - (b\Delta\xi)\Lambda c - (2\Delta\xi)Q - (2\Delta\xi)s'_t.$$

The left-hand side is the time rate of change of the mass in the section; the first term on the right-hand side is the net flux through the cross sections at ξ and $\xi + \Delta\xi$ (ϕ is the flux); the second term is the decay rate and the third term is rate of loss of contaminant to the porous blocks through the upper and lower faces $[Q = Q(\xi, \tau)$ is measured in mass per unit area per unit time]; the last term is the rate of adsorption to the fracture walls, with $s' = s'(\xi, \eta, \tau)$ given in mass of contaminant per unit area. We assume linear adsorption, i.e., $s' = K'_d c$, and we assume that the loss term is proportional to the concentration gradient in the porous block, i.e., we assume the constitutive relation

$$Q = -D''\omega m_\eta(\xi, 0, \tau).$$

The constitutive relation for the flux ϕ consists of a dispersive flux term and a convection term; that is,

$$\phi = -Dc_\xi + Vc,$$

where D is the constant dispersivity and V is the Darcy velocity. Upon dividing the mass balance law by $b\Delta\xi$, taking the limit as $\Delta\xi$ goes to zero, and using the definitions above, we obtain the mass balance equation in the fracture channel in the form

$$Rc_\tau = Dc_{\xi\xi} - Vc_\xi - \Lambda c + \frac{2\omega D''}{b}m_\eta(\xi, 0, \tau), \tag{2.89}$$

where R is a surface retardation factor in the fracture given by $R = 1 + 2K'_d/b$. Note that our assumption is that the fracture is a channel; that is, we have chosen the porosity to be unity.

Along the interface $\eta = 0$ between the fracture and the porous block we assume that the concentration is continuous, i.e.,

$$m(\xi, 0, \tau) = c(\xi, \tau).$$

Thus, also, $m(\xi, -b, \tau) = c(\xi, \tau)$ along the lower fracture interface.

There are several characteristic times that could be used to rescale the problem and reduce it to dimensionless form. For example, there are time scales for convection, dfor iffusion, and for the rate that contaminant is being injected at the input boundary. We shall scale time on the basis of convection, which is reasonable because the intended observations will often involve measuring contaminant output from the fracture at some distance downstream. Let L be a length scale in the ξ direction, and let $T = L/V$ be the time scale. We take the η length scale to be $Y = \sqrt{D''T}$. For simplicity we assume the retardation parameters to be unity, i.e., $R = R' = 1$. Then, a change to dimensionless variables x, y, and t defined by

$$x = \frac{\xi}{L}, \quad y = \frac{\eta}{Y}, \quad t = \frac{\tau}{T},$$

transforms (2.88) and (2.89) into

$$
\begin{aligned}
m_t &= \epsilon m_{xx} + m_{yy} - \lambda m, & (2.90) \\
c_t &= \alpha c_{xx} - c_x - \lambda c + \gamma m_y(x, 0, t), & (2.91)
\end{aligned}
$$

where

$$\epsilon \equiv \frac{D'}{VL}, \quad \alpha \equiv \frac{D}{VL},$$

and

$$\gamma \equiv \frac{2\omega\sqrt{LD''/V}}{b}.$$

Here, $\lambda = L\Lambda/V$. At the inlet boundary of the fracture we assume a Dirichlet condition

$$c(0, t) = c_0(t), \quad t > 0,$$

where c_0 is a given continuous function representing an input concentration, and $c_0(0) = 0$. Initially we assume the absence of contaminant, or

$$m(x, y, 0) = 0, \quad x, y \geq 0; \quad c(x, 0) = 0, \quad x \geq 0.$$

Along the boundary $x = 0$ we impose a no-flux condition

$$m_x(0, y, t) = 0, \quad y, t > 0. \qquad (2.92)$$

Our basic assumptions to simplify the model are

$$\epsilon \ll 1, \quad \alpha \ll 1, \quad \gamma = 0(1).$$

So, on a convection time scale, diffusion in the ξ direction in the porous block is assumed to be small. Furthermore, the smallness of α imposes the condition that dispersion can be neglected in the fracture; such an assumption is valid when the advective flux in the fracture is large [Sudicky and Frind (1982)].

Therefore, the simplified model equations ($\alpha = \epsilon = 0$) become

$$m_t = m_{yy} - \lambda m, \tag{2.93}$$
$$c_t = -c_x - \lambda c + \gamma m_y(x, 0, t). \tag{2.94}$$

We now observe that, because $m(x, 0, t) = c(x, t)$, (2.94) can be regarded as an evolutionary boundary condition on m along the fracture interface $y = 0$. Thus, we write (2.94) as

$$m_t(x, 0, t) = -m_x(x, 0, t) - \lambda m(x, 0, t) + \gamma m_y(x, 0, t).$$

Furthermore, one can remove decay from the preceding equations by the transformation $u(x, y, t) = m(x, y, t)e^{\lambda t}$. Then the final form of the boundary value problem becomes

$$u_t = u_{yy}, \quad x, y > 0, \quad t > 0, \tag{2.95}$$
$$u_t + u_x = \gamma u_y, \quad x, t > 0, \quad y = 0, \tag{2.96}$$
$$u(0, 0, t) = u_0(t), \quad t > 0, \tag{2.97}$$
$$u(x, y, 0) = 0, \quad x, y \geq 0, \tag{2.98}$$

where $u_0 \equiv c_0 e^{\lambda t}$.

Observe that the no-flux boundary condition (2.92) has not been included in the problem (2.95)–(2.98). Indeed, neglecting diffusion in the porous block in the x direction precludes a boundary condition along $x = 0$; otherwise we obtain an ill-posed problem. This means that, in the simplified model, there can be contaminant flux along the boundary $x = 0$. [Logan, Ledder, and Homp (1998) have shown that this problem (2.95)–(2.98) is the leading order problem corresponding to a singular perturbation problem (2.90)–(2.91) with $\alpha = O(\sqrt{\epsilon})$, with a boundary layer in the gradient along $x = 0$.]

Using Laplace transforms we now obtain the solution to (2.95)–(2.98). We proceed formally. Let $\psi(x, t)$ denote the (unknown) value of the concentration u along $y = 0$, and consider the following boundary value problem in y and t:

$$u_t = u_{yy}, \quad y > 0, \quad t > 0, \tag{2.99}$$
$$u(x, 0, t) = \psi(x, t), \quad t > 0, \tag{2.100}$$
$$u(x, y, 0) = 0, \quad y > 0, \tag{2.101}$$

where $x > 0$ is regarded as a parameter. From the results in Chapter 1 we know that the solution to (2.99)–(2.101) is given by the convolution integral

$$u(x, y, t) = h(y, \cdot) \star \psi(x, \cdot) = \int_0^t h(y, t - s)\psi(x, s)\,ds, \tag{2.102}$$

where h is given by

$$h(y, t) = \frac{ye^{-y^2/4t}}{\sqrt{4\pi t^3}}.$$

If $U(x, y, p)$ and $\Psi(x, p)$ denote the Laplace transforms (on t) of u and ψ, respectively (here, we use the symbol p is the transform variable, rather than s), then applying the convolution theorem to (2.102) gives

$$U(x, y, p) = e^{-y\sqrt{p}}\Psi(x, p).\tag{2.103}$$

Now take the Laplace transform of (2.96) to obtain

$$pU(x, 0, p) = -U_x(x, 0, p) + \gamma U_y(x, 0, p),$$

or

$$\Psi_x(x, p) = -(p + \gamma\sqrt{p})\Psi(x, p),$$

which is a differential equation for Ψ. From condition (2.97) we obtain $\Psi(0, p) = U_0(p)$, where U_0 is the transform of u_0. Thus,

$$\Psi(x, p) = U_0(p)e^{-(p+\gamma\sqrt{p})x}.$$

This gives, from (2.103),

$$U(x, y, p) = U_0(p)e^{-(p+\gamma\sqrt{p})x - y\sqrt{p}},$$

which is the solution to the problem in the transform domain. By the convolution theorem we then obtain

$$u(x, y, t) = u_0(t) \star \frac{H(t - x)(\gamma x + y)}{\sqrt{4\pi}(t - x)^{3/2}}e^{-(\gamma x + y)^2/4(t - x)},\tag{2.104}$$

where H is the Heaviside function. Here we have used the well-known transform [see, for example, Carslaw and Jaeger (1959)]

$$\mathcal{L}\left(\frac{z}{\sqrt{4\pi t^{3/2}}}e^{-z^2/4t}\right) = e^{-z\sqrt{p}},$$

and the shift theorem

$$\mathcal{L}(H(t - z)f(t - z)) = F(p)e^{-pz}.$$

Equation (2.104) can be written concisely as

$$u(x, y, t) = \int_0^{t-x} h(\gamma x + y, t - s - x)u_0(s)ds,$$

which is the solution to (2.95)–(2.98).

Exercise 48 *Follow the ideas presented in this section and formulate a model of flow on the domain $x > 0$ through an infinite system of parallel fractures (parallel to the x-axis), each of width b and each separated by a porous block of width h. Assume zero initial conditions and a periodic boundary condition on the solute concentration at the inlets to the fractures along $x = 0$. Reduce the model to dimensionless form, neglect dispersion in the porous blocks in the x-direction and obtain a solution for the problem. Logan, Zlotnik, and Cohn (1996) can be consulted for details.*

2.8 Reference Notes

There are several elementary, hydrogeology books that have a mathematical flavor, for example, de Marsily (1986) and Fetter (1993). Books that discuss mathematical modeling from an advanced viewpoint are Fowler (1997), Lin and Segel (1989), and Logan (1987, 1997c). Accessible texts for the underlying mathematical fluid mechanics are Acheson (1990), Chorin and Marsden (1993), and Bird, Stewart, and Lightfoot (1960). There are many reference texts for nonlinear PDEs; books dealing strictly with the subject are Logan (1994), Smoller (1994), and Whitham (1974). The book by Knobel (2000) is an excellent introduction to nonlinear waves. Gelhar (1993) has an introduction to the stochastic approach to groundwater problems.

Chapter 3

Traveling Wave Solutions

In this chapter we study model problems that lead to an important, special class of solutions called traveling wave solutions. Examining the behavior of these solutions can give insights into the role of competing mechanisms in a given problem, for example, reaction versus dispersion or advection versus dispersion.

3.1 Examples of Traveling Waves

In laboratory tracer experiments in long, packed columns, researchers have observed that upon applying a constant inlet tracer concentration at one end, under some conditions there resulted a concentration wave progressing through the column moving at constant speed and maintaining the same shape. These concentration waves have the mathematical form $u(x, t) = U(z)$, where $z = x - ct$, and they are called **traveling waves**. See figure 3.1. Here $U(z)$ is the wave form and c is the wave speed. If U is a smooth function and approaches constant values at $\pm\infty$, then the traveling wave is called a **wave front**. If the constant states at $\pm\infty$ are equal, the traveling wave is called a **pulse**. Most of the wave front profiles we encounter are monotone, i.e., they either increase or decrease. We remark that wave fronts are solutions defined on the entire real line and represent only a mathematical model of actual wave profiles on finite domains in the laboratory.

The simplest equation that admits wave front solutions is the pure advection equation

$$u_t = -cu_x.$$

In this case *all* solutions are of the form of a right traveling wave $u = U(x - ct)$, where U is an arbitrary function; the actual form of U is determined from initial or boundary conditions. It is easy to see, however, that the equilibrium model without dispersion,

$$(u + \beta f(u))_t = -u_x,$$

does not admit nonconstant wave front solutions; neither does the dispersion

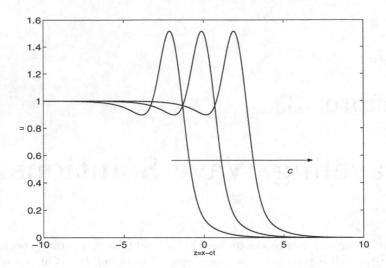

Figure 3.1: A traveling wave pictured as a moving wave profile in spacetime.

equation

$$u_t = \alpha u_{xx},$$

nor the advection–dispersion equation

$$u_t = \alpha u_{xx} - u_x.$$

Wave front solutions do exist for certain nonlinear parabolic problems when there is a balance of competing processes like dispersion, adsorption, and advection. For example, the equilibrium model with dispersion

$$(u + \beta f(u))_t = \alpha u_{xx} - u_x$$

admits wave front solutions under special conditions.   We shall discuss this problem in detail in the first section below.

We point out that wave front solutions must approach constant, or equilibrium, states at infinity. Plane wave solutions of the form

$$u = Ae^{ik(x-ct)}$$

are traveling waves, but they are not wave fronts because of the oscillations of the real and imaginary parts of u.

Exercise 49 *Verify that the dispersion equation and the advection–dispersion equation given above do not admit nonconstant traveling wave solutions [Hint: substitute $u = U(x - ct)$ into the equation and solve for U, determining the constants of integration from the constant conditions at infinity.]*

This chapter deals with the existence of wave front solutions, and their stability (or permanence), for several mathematical models involving advection, dispersion, and adsorption. This important class of solutions has become fundamental in the theory of nonlinear parabolic systems. The analysis of traveling wave solutions is the simplest way to study the interactions among different types of processes.

The idea can be explained in a simple way. Consider the parabolic equation

$$u_t = \alpha u_{xx} + f(u, u_x).$$

Here, α is a dispersion coefficient and the function f contains advection and reaction terms. We assume $u = U(z)$, where $z = x - ct$, and where both the wave speed c and the shape of the wave form U are to be determined. We think of z as a moving coordinate. By the chain rule we can compute the partial derivatives that occur in the equation:

$$u_t = -cU'(z), \quad u_x = U'(z), \quad u_{xx} = U''(z),$$

where "prime" denotes the ordinary derivative d/dz. Thus, substituting the unknown wave form solution into the partial differential equation gives

$$-cU' = \alpha U'' + f(U, U'), \tag{3.1}$$

which is a second-order, ordinary differential equation for the wave form $U(z)$. Because we want wave front solutions that have constant states at plus and minus inifinity, we have boundary conditions of the form

$$U(-\infty) = u_l, \quad U(+\infty) = u_r, \tag{3.2}$$

where u_l and u_r are the states at infinity. If $u_l = u_r$, we refer to the wave form as a **pulse**. If $u_l > u_r$, the wave is a **compression wave,** and if $u_l < u_r$, it is called a **rarefaction wave,** or expansion wave. If the subject is about the transport of a contaminant solute, compression waves are often called **contamination waves** and rarefactions are **remediation waves**. In summary, the problem of finding wave front solutions is reduced to the problem of solving the boundary value problem (3.1)–(3.2) for U on $\mathbb{R}$. Because the solution $U(z)$ is a function of a single variable z that depends on x and t, it is an example of a similarity solution.

Showing the existence of a solution to a boundary value problem for an ordinary differential equation can be a formidable task. Sometimes the second-order equation can be attacked directly. A different strategy is to introduce the phase coordinate $V = U'$ and work in the UV phase plane. There the ordinary differential equation (3.1) reduces to a first-order autonomous system

$$\begin{aligned} U' &= V \\ V' &= -\alpha^{-1}(cV + f(U, V)). \end{aligned}$$

Existence is then deduced by demonstrating there is a unique phase trajectory, or orbit, connecting two critical points, the latter representing the equilibrium states at minus and plus infinity: $(u_l, 0)$ and $(u_r, 0)$. Because the differential equations are autonomous (independent of z), then any horizontal translate of a solution is again a solution. That is, if $u = U(z)$ represents a traveling wave, then so does $u = U(z + z_0)$ for any constant z_0. This means that we can impose another condition to uniquely select a solution; usually we force the wave to have a certain value at $z = 0$ by imposing a condition $U(0) = u_0$ for some u_0.

We now illustrate the procedure for finding traveling waves with some simple examples.

Consider the nonlinear advection–dispersion equation

$$u_t = \alpha u_{xx} - g(u)_x, \tag{3.3}$$

where $g(u)$ is a given function. If $g(u) = u$, then we obtain the linear advection–dispersion equation, which, as mentioned above, has no nonconstant traveling wave solution. The required assumption on the function g to ensure the existence of traveling waves of (3.3) is strict convexity. Therefore, we assume that g is defined and twice continuously differentiable on all of $\mathbb{R}$ and

$$g''(u) > 0 \quad \text{or} \quad g''(u) < 0.$$

Then the PDE has a fully nonlinear advection term. We assume $u = U(z)$, where $z = x - ct$, and

$$U(-\infty) = u_l, \quad U(+\infty) = u_r.$$

We obtain

$$-cU' = \alpha U'' - g(U)'.$$

Integrating once gives

$$-cU = \alpha U' - g(U) + b,$$

where b is a constant of integration. Applying the boundary condition at $z = +\infty$ in (3.4) gives

$$b = g(u_r) - cu_r.$$

Applying the boundary condition at $-\infty$ then gives $-cu_l = -g(u_l) + g(u_r) - cu_r$. Therefore, if a traveling wave solution exists, then the wave speed c is given by

$$c = \frac{g(u_l) - g(u_r)}{u_l - u_r}. \tag{3.4}$$

To finally determine the traveling wave profile we write the differential equation in the form

$$\alpha U' = g(U) - cU - b.$$

Separating variables gives

$$\frac{\alpha dU}{g(U) - cU - b} = dz.$$

Integrating from $z = 0$ to an arbitrary z, while U varies from $U(0) = d$ to an arbitrary U, yields

$$\alpha \int_d^U \frac{dw}{g(w) - cw - b} = z. \tag{3.5}$$

Equation (3.5) defines the traveling wave $U = U(z)$ implicitly. Therefore, if g satisfies the strict convexity property and u_l and u_r are any two constants, there exists a unique traveling wave solution connecting these two states, where the wave speed c is given by (3.4).

Example 50 *If we take $g(u) = \frac{1}{2}u^2$ in (3.3), then we obtain the classical **Burgers' equation***

$$u_t = \alpha u_{xx} - u u_x.$$

If $u_l = 1$ and $u_r = 0$, then the wave speed is $c = \frac{1}{2}$ and $b = 0$. Then from (3.5) we have

$$2\alpha \int_{1/2}^U \frac{dw}{w^2 - w} = z,$$

where we have imposed $U(0) = 0.5$. Integrating and solving for U gives the wave profile

$$U(z) = \frac{1}{1 + e^{z/2\alpha}}.$$

Therefore, traveling wave solutions are given by

$$u(x, t) = \frac{1}{1 + e^{(x-ct)/2\alpha}}.$$

Exercise 51 *Show that the equation*

$$u_t = u_{xx} - u^2 u_x,$$

*admits both **compression** ($u_l > u_r$) and **rarefaction** ($u_l < u_r$) wave fronts.*

In the preceding examples we were able solve the ordinary differential equation for the wave profile $U(z)$ to find traveling waves directly. Now we examine a problem where a phase plane argument is useful. The nonlinear reaction–dispersion equation

$$u_t = u_{xx} + f(u)$$

has been examined extensively in the literature [see, for example, Volpert and Volpert (1993) for a large bibliography]. The prototypes are **Fisher's equation**

$$u_t = \alpha u_{xx} + u(1 - u),$$

which has logistic growth, and the **bistable model**

$$u_t = \alpha u_{xx} + u(u - a)(b - u), \quad a, b > 0.$$

For the moment let us make no assumptions regarding the form of the reaction term $f(u)$, other than it is continuously differentiable on $\mathbb{R}$. We assume $u = U(z)$, where $z = x - ct$, and

$$U(-\infty) = u_l, \quad U(+\infty) = u_r.$$

Then we obtain the differential equation for the wave profile as

$$-cU' = U'' - f(U).$$

This equation cannot be integrated directly. So, we introduce $V = U'$ and write the equation as a first-order system, or two-dimensional dynamical system, as

$$\begin{aligned} U' &= V, \\ V' &= -f(U) - cV. \end{aligned}$$

If a traveling wave exists, then $U(z)$ must approach constant states as $z \to \pm\infty$, and $U'(z) = V(z)$ must approach zero. Therefore, the states (U, V) at minus and plus infinity must be critical points of the system. These states are $(u_l, 0)$ and $(u_r, 0)$. Therefore, f must have zeros at the two values u_l and u_r, i.e., $f(u_l) = f(u_r) = 0$. In that case, depending upon the overall structure of f, we may be able to prove for certain values of c that there exists a unique trajectory or orbit of the system that connects the two points $(u_l, 0)$ and $(u_r, 0)$. This orbit, which is called a **heteroclinic orbit** if $u_l \neq u_r$, would then represent the traveling wave solution we seek. The wave speed c can be thought of as an eigenvalue and the problem itself a nonlinear eigenvalue problem; the wave front is the corresponding eigenfunction.

To fix the idea let us take $f(u) = u(1 - u)$ and $\alpha = 1$ and consider

$$u_t = u_{xx} + u(1 - u),$$

which is Fisher's equation. Then the dynamical system is

$$\begin{aligned} U' &= V, & (3.6) \\ V' &= -U(1 - U) - cV, & (3.7) \end{aligned}$$

which has two critical points, $(0, 0)$ and $(1, 0)$, in the UV plane. The overall phase portrait depends on the parameter c. The question is: Do there exist values of the wave speed c such that there is a heteroclinic orbit connecting the two critical points? The answer to this question is more difficult than it may first appear, and it depends on the

local behavior of the solution near the critical points. Figures 3.2 and 3.3 show phase portraits in the two cases $c = 1$ and $c = 4$. Note in both cases that $(1, 0)$ is a saddle point, and that in the case $c = 1$ the origin is a stable spiral, while in the case $c = 4$ it is a stable node. In both cases there appears to be a heteroclinic orbit, representing a traveling wave, connecting $(1, 0)$ to $(0, 0)$ as z goes from $z = -\infty$ to $z = +\infty$. If we consider, for physical reasons, only positive values of the density U, then we would reject the case $c = 1$.

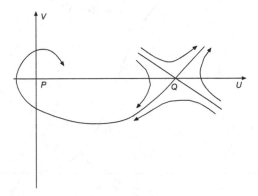

Figure 3.2: Phase portrait for (3.6)–(3.7) when $c = 1$.

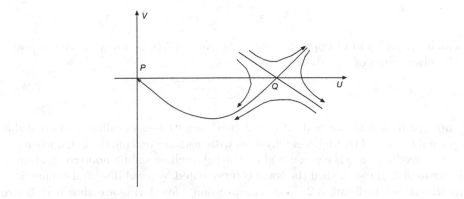

Figure 3.3: Phase portraits for (3.6)–(3.7) when $c = 4$.

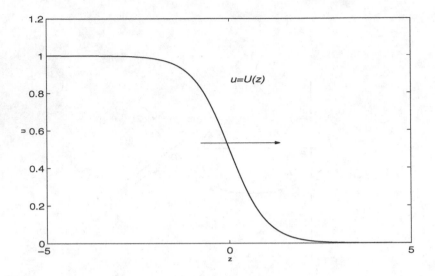

Figure 3.4: Traveling wave for (1.6)–(1.7) when $c = 4$.

A more careful analysis can be given by looking at the local structure near the critical points. The Jacobian matrix for the system is

$$J(U, V) = \begin{pmatrix} 0 & 1 \\ 2U - 1 & -c \end{pmatrix}. \tag{3.8}$$

It is easy to check that the eigenvalues of $J(1, 0)$ are

$$\lambda_\pm = \frac{-c \pm \sqrt{c^2 + 4}}{2},$$

which are real and of opposite sign. Therefore, $(1, 0)$ is always a saddle point. The eigenvalues of $J(0, 0)$ are

$$\lambda_\pm = \frac{-c \pm \sqrt{c^2 - 4}}{2}.$$

Thus, $(0, 0)$ is a stable node if $c \geq 2$ (real, negative eigenvalues), and a stable spiral if $0 < c < 2$ (complex eigenvalues with negative real part). Thus, when $c \geq 2$ the traveling wave is represented by the heteroclinic saddle-node connection as in figure 3.3. If $c < 2$, then the wave is represented by a saddle-spiral connections as illustrated in figure 3.2. The compression wave forms are shown in figure 3.4. The actual phase diagram, or phase portrait, picturing the various orbits can be obtained using a computer algebra package, e.g., MATLAB, Maple, or Mathematica. The proof that a desired orbit exists often is difficult and requires careful methods of mathematical analysis.

Remark 52 *There is an interesting connection between traveling waves for reaction–dispersion equations and parabolic equations with nonlinear dispersion.*

The result of Engler (1985) is that if $f(r) = g(r)D(r)$, $f(0) = g(0) = f(1) = g(1) = 0$, and $D(r) > 0$, then a traveling wave solution to the equation

$$u_t = (D(u)u_x)_x + g(u)$$

exists if, and only if, a traveling wave solution to the reaction–dispersion equation

$$v_t = v_{xx} + f(v)$$

exists. See Grindrod (1996, p 226) for an elementary discussion of the transformation involved in showing this equivalence.

Exercise 53 *Show that the Korteweg–de Vries (KdV) equation*

$$u_t = -ku_{xxx} - u\,u_x$$

admits a traveling wave that is a pulse (called a soliton). This traveling wave is called a soliton and has the form

$$u = U(z) = \gamma + (\alpha - \gamma)\sec \mathrm{h}^2(\sqrt{\frac{\alpha - \gamma}{12k}}z),$$

where γ is the state at $\pm\infty$, and $\alpha - \gamma$ is the amplitude of the pulse. Note that the wave speed depends upon the amplitude. Show that the wave speed is $c = (\alpha + 2\gamma)/3$. The KdV equation arises in many contexts, one being a model of special surface waves on water [see Drazin and Johnson (1989)].

3.2 Hydrogeological Models

3.2.1 The Equilibrium Model

We now begin to examine the question of existence of traveling waves to some of the hydrogeological models developed in Chapter 2. We first study the equilibrium equation

$$(1 + \beta f'(u))u_t = \alpha u_{xx} - u_x, \tag{3.9}$$

where the isotherm f satisfies the conditions

$$f(0) = 0, \quad f > 0, \quad f' > 0, \quad f'' < 0 \text{ for } u > 0,$$

and $\alpha, \beta > 0$. We will show the existence of traveling wave solutions of the form $u = U(z)$, $z = x - ct$, for some positive wave speed c, where $U \in C^2(\mathbb{R})$ and satisfies the boundary conditions

$$U(-\infty) = u_l > 0, \ U(\infty) = 0. \tag{3.10}$$

We record the result formally.

Theorem 54 *Let $f \in C^2(\mathbb{R})$ with $f(0) = 0$, and $f > 0$, $f' > 0$, $f'' < 0$ for $U > 0$. Then there exists a unique, monotonically decreasing, traveling wave solution $u = U(z)$, $z = x - ct$, to (3.9) satisfying the boundary conditions (3.10) and having wave speed c given by*

$$c = \frac{u_l}{u_l + \beta f(u_l)}.$$

Proof. *We present the key steps, which are straightforward to verify. Substituting the assumed wave form into (3.9) leads to the ordinary differential equation for U, namely,*

$$-c(1 + \beta f'(U))U' = \alpha U'' - U'.$$

This equation integrates directly to give the first-order equation

$$U'(z) = \alpha^{-1}((1 - c)U - \beta c f(U)) \equiv G(U), \tag{3.11}$$

where the boundary condition at plus infinity was used to evaluate the constant of integration (making it zero). We seek a solution on $\mathbb{R}$ that connects two equilibrium states $U = u_l$ and $U = 0$; the former must be unstable, and the latter must be asymptotically stable. Clearly, $U = 0$ is an equilibrium solution; in order that $U = u_l$ be an equilibrium state, we must have $G(u_l) = 0$, or, equivalently,

$$c = \frac{u_l}{1 + \beta f(u_l)},$$

which fixes the wave speed c, $0 < c < 1$. Note that $G_{UU}(U) = -\alpha^{-1}\beta f''(U) < 0$ for $U > 0$. Thus the graph of G is (strictly) concave upward on the interval $(0, u_l)$. Because G vanishes at $U = 0, u_l$, it follows that $G_U(0) < 0$ and $G_U(u_l) > 0$. Thus, (3.11) has an attractor at $U = 0$ and a repeller at $U = u_l$, which completes the proof. ∎

As the following exercises show, for some isotherms the traveling wave solution can be determined explicitly.

Exercise 55 *Consider the quadratic isotherm defined by*

$$f(u) = bu - \gamma u^2, \quad b > \gamma.$$

In this case show that the differential equation (3.11) is a Bernoulli equation, which has solution

$$U(z) = \frac{1}{1 + e^{az}}, \quad a \equiv b\gamma c\alpha,$$

where we have taken $U(0) = 0.5$. The wave speed is then $c = (1 + b - \beta\gamma)^{-1}$.

Exercise 56 *In the case of the Langmuir isotherm*

$$f(u) = \frac{u}{1 + au},$$

show that the wave speed c is given by

$$c = \frac{1 + au_l}{1 + \beta + au_l},$$

and the traveling wave is given implicitly by

$$\alpha \int_{u_l/2}^{U} \frac{1 + aw}{w((1-c)(1+au) - \beta c)} dw = z.$$

In the case that the isotherm is a Freundlich isotherm,

$$f(u) = \sqrt{u},$$

the analysis is slightly different. The equilibrium model becomes

$$(u + \beta\sqrt{u})_t = \alpha u_{xx} - u_x.$$

Proceeding with the Ansatz of a traveling wave solution, we obtain the ordinary differential equation

$$\alpha U' = (1 - c)U - c\beta\sqrt{U},$$

where the wave speed is

$$c = \frac{u_l}{u_l + \beta\sqrt{u_l}} < 1.$$

Observe that

$$\frac{c\beta}{1 - c} = \sqrt{u_l}.$$

The differential equation is a Bernoulli equation that can be reduced to a linear equation with the substitution $W = \sqrt{U}$. We obtain

$$W' = \frac{1 - c}{2\alpha} W - \frac{c\beta}{2\alpha}.$$

Thus,

$$W = \sqrt{U} = ke^{(1-c)z/2\alpha} - \frac{c\beta}{1 - c},$$

where k is a constant of integration. Therefore,

$$U(z) = \left(ke^{(1-c)z/2\alpha} - \sqrt{u_l}\right)^2.$$

As $z \to -\infty$ we satisfy the boundary condition $U(-\infty) = u_l$. But, as well, $U(z) \to 0$ for a finite value of z. Choosing this finite value of z to be zero, that is, $U(0) = 0$, we get $k = \sqrt{u_l}$. Hence, we have the traveling wave profile

$$U(z) = u_l \left(e^{(1-c)z/2\alpha} - 1\right)^2, \quad z < 0; \quad U(z) = 0, \quad z \geq 0.$$

3.2.2   The Nonequilibrium Model

Now we investigate the nonequilibrium model

$$u_t = \alpha u_{xx} - u_x - \beta s_t, \tag{3.12}$$
$$s_t = F(u,s) = -s + f(u), \tag{3.13}$$

where the assumptions on f are, as before, $f(0) = 0, f'(u) > 0, f''(u) < 0$ for $u > 0$. Again $\alpha, \beta > 0$. We look for wave front solutions of the form

$$u = U(z), \quad s = S(z), \quad z = x - ct,$$

with

$$U(+\infty) = S(+\infty) = 0, \quad U(-\infty) = 1.$$

Observe that $S(-\infty)$ is fixed by the equilibrium condition

$$S(-\infty) = f(U(-\infty)) = f(1).$$

It immediately follows that U and S must satisfy the dynamical system

$$U' = \alpha^{-1}((1-c)U - c\beta S), \tag{3.14}$$
$$S' = -c^{-1}(-S + f(U)). \tag{3.15}$$

Clearly $U = 0, S = 0$ is a critical point, and it represents the equilibrium state at $z = +\infty$. For $U = 1, S = f(1)$ to be an equilibrium state at $z = -\infty$ we must have $1 - c - c\beta f(1) = 0$, or

$$c = \frac{1}{1 + \beta f(1)}, \tag{3.16}$$

which uniquely determines the wave speed c.

To analyze the behavior of trajectories in the phase plane it is often beneficial to sketch the **nullclines** $U' = 0, S' = 0$. The nullclines are the loci of points in the phase plane where the vector field defined by the system are vertical or horizontal. The nullclines $U' = 0, S' = 0$ intersect at the critical points, and in the present problem at the point $(1, f(1))$ (see figure 3.5). A traveling wave solution is guaranteed by the existence of a unique heteroclinic orbit connecting the unstable node $(1, f(1))$ and the saddle point $(0,0)$. We note that horizontal vector field points left along the nullcline $S' = 0$, and it points vertically downward along the nullcline $U' = 0$ Thus, the boundary of the region in between the two nullclines consists entirely of egress points (i.e., where the vector field is exiting). Therefore, there is a unique orbit connecting the two critical points [see Hartman (1965)]. It is easy to see that the critical point $(0,0)$ is a saddle point, and $(1, f(1))$ is an unstable node. Therefore, the orbit representing the traveling wave is a node–saddle connection. It is clear that both U and S are monotonically decreasing.

It is straightforward to eliminate S from equations (3.14)–(3.15) to obtain a single equation for the wave front U. We get

$$\alpha c U'' + (c^2 - c - \alpha)U' + (1 - c)U = c\beta f(U).$$

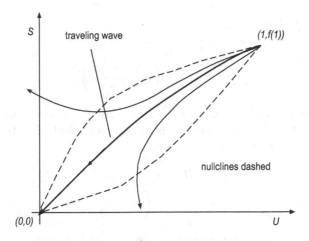

Figure 3.5: Phase portrait for (3.14)–(3.15).

Here α is the inverse of a Peclet number, the latter measuring the ratio of advection to dispersion. Thus, for small α, advection dominates dispersion. In this limit one can obtain an approximate solution to the second-order differential equation. Setting $\alpha = 0$ gives the van der Zee (1990) approximation

$$(c^2 - c - \alpha)U' + (1 - c)U = c\beta f(U). \tag{3.17}$$

Separating variables and integrating gives an implicit form of the solution, namely,

$$\int_a^U \frac{dw}{\frac{\beta}{c-1}f(w) + \frac{1}{c}w} = z.$$

Of course, as $\alpha \to 0$ the second-order differential equation (3.17) is singular; the validity of the van der Zee approximation, using singular perturbation methods, is discussed in Logan and Ledder (1995).

3.2.3 A Nonlinear Advection Model

We showed in Section 3.2 that the Burgers' equation

$$u_t = \alpha u_{xx} - uu_x$$

admits traveling waves. For Burgers' equation the effect of the nonlinear advection term uu_x is to "shock up" wave profiles, causing a gradually steppening of wave; at the same time the dispersion term u_{xx} causes the wave to smear out. In the traveling wave these two competing effects balance and a fixed wave form can be propagated.

In the next few paragraphs we examine a nonequilibrium hydrogeological model with both nonlinear advection and adsorption, as well as dispersion.

Consider the system

$$u_t = u_{xx} - uu_x - \beta s_t, \tag{3.18}$$

$$s_t = -k\left(s - \frac{u^2}{1+u^2}\right). \tag{3.19}$$

We seek a solution of the form $u = U(z), s = S(z), z = x - ct$ that vanishes at $z = +\infty$ and $U(-\infty) = u_1$, where u_1 is to be determined. Substituting into the system of equations (3.18)–(3.19) gives

$$U'' = (U - c)U' - \beta cS',$$

$$S' = -kc^{-1}\left(S - \frac{U^2}{1+U^2}\right).$$

The first equation can be integrated immediately and the constant of integration can be determined from the boundary condition at $+\infty$. We then obtain the dynamical system

$$U' = \frac{U^2}{2} - cU - \beta cS, \tag{3.20}$$

$$S' = -kc^{-1}\left(S - \frac{U^2}{1+U^2}\right). \tag{3.21}$$

The nullclines, where $U' = 0$ and $S' = 0$, are given by the parabola

$$S = 2(\beta c)^{-1}U(U - 2c),$$

and the equilibrium curve

$$S = \frac{U^2}{1+U^2},$$

respectively. The phase plane is shown in figure 3.6. The origin is clearly a critical point representing the state at plus-infinity. Other critical points must occur as intersections of the nullclines. Eliminating S from the algebraic system, we obtain the cubic equation

$$U^3 - 2cU^2 + (1 - 2\beta c)U - 2c = 0. \tag{3.22}$$

The positive real roots of this equation give the U-coordinate of other critical points in the first quadrant. It is clear that, if $c > 0$, then there is at least one positive real root, and we denote the smallest positive real root by u_1. Thus, (u_1, s_1) is a critical point, where $s_1 = u_1^2/(1 + u_1^2)$. From (3.21) we then obtain the wave speed $c = 0.5u_1^2/(u_1 + \beta s_1) > 0$. Figure 3.6 shows the case of one positive root of the cubic.

In this case the parabola lies below the equilibrium curve. It is easy to check that the Jacobi matrix $J(U, S)$ has real eigenvalues of opposite sign ($-c$ and k/c) at $(0, 0)$; thus, the origin is a saddle point. The eigenvector associated with the negative eigenvalue is $(1, 0)^t$ (here, the superscript t denotes transpose),

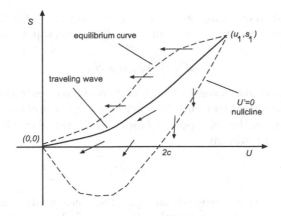

Figure 3.6: Phase portrait for (3.20)–(3.21).

and therefore the stable manifold W_s enters the origin tangent to the U axis. Because of the structure of the vector field along the nullclines and along $U = 0$, it is observed that there is a unique heteroclinic orbit to (3.20)–(3.21), the separatrix, that connects the critical point (u_1, s_1) to the origin as z varies from $-\infty$ to $+\infty$. Consequently, we may record the following result.

Theorem 57 *Consider the nonequilibrium model given by (3.18)–(3.19) where k and β are positive constants. For any $c > 0$ there exists a smooth, positive, decreasing, traveling wave solution $u = U(x - ct)$, $s = S(x - ct)$, with the properties $U(+\infty) = S(+\infty) = 0$ and $U(-\infty) = u_1$, $S(-\infty) = s_1$, where u_1 is the smallest positive root of the cubic equation (3.22).*

Further details involved in this calculation can be found in Cohn and Logan (1995a).

3.3 Discontinuous Wave Fronts

As we have observed, wave front solutions arise in a variety of physical and biological problems, and their study provides a simple way to examine interactions among advection, dispersion, and reaction processes. Now we examine a solute transport model in a hydrogeological setting where the dispersive portion of the flux remains bounded even when the gradients become large. An analysis of the effects of bounded dissipation in Burgers-type equations can be found in Kurganov and Rosenau (1997, 1998). By including adsorption phenomena in the model we find interesting interplay among the physical processes of advection, dispersion, and the kinetics of adsorption. One such result is that discontinuous wave fronts, or shocks, can propagate, as well as continuous fronts. Which case actually occurs depends upon the dominance of the various processes in the system. The present discussion is adapted from Homp and Logan (1999).

The standard mass balance equation governing the advection, dispersion, and adsorption of a chemical tracer in a one-dimensional porous domain is (see, Section 2.2)

$$\omega C_t = -\phi_x - \rho_b S_t,$$

where ω is the constant porosity, $C = C(x,t)$ is the concentration of the solute, $S = S(x,t)$ is the sorbed concentration per unit mass of solid, and $\rho_b = (1-\omega)\rho$ is the bulk density of the solid porous fabric. The kinetics of adsorption is given by the equilibrium isotherm

$$S = F(C/C_0),$$

where C_0 is a reference concentration, e.g., the concentration at the inlet. We recall that the constitutive assumption for the flux Q was previously given by

$$Q = -\omega D C_x + VC,$$

where D is the dispersion coefficient and V is the Darcy velocity. That is, the flux consists of advective portion VC, caused by the bulk motion of the liquid, and a dispersive flux $-\omega D C_x$ proportional to the concentration gradient. We recall that D contains both a molecular diffusion component and kinematic dispersion component due to the flow through the tortuous pathways in the fabric. These equations lead to the equilibrium model

$$C_t = D C_{xx} - v C_x - \frac{\rho_b}{\omega} F(C/C_0)_t,$$

where $v = V/\omega$ is the average velocity.

Because the flux is assumed to be linear in the gradient, the response of the system to a steep concentration front may become unphysical because the flux becomes unbounded. Linearity in the gradient, or Fick's law, can be regarded as resulting from a Taylor expansion and an assumption of small gradients. For example, if $Q = Q(u_x)$ is the flux, then

$$
\begin{aligned}
Q(u_x) &= Q(0) + Q'(0)u_x + \frac{1}{2}Q''(0)u_x^2 + \frac{1}{3!}Q'''(0)u_x^3 + \cdots \\
&= Q'(0)u_x + \frac{1}{3!}Q'''(0)u_x^3 + \cdots \\
&\approx Q'(0)u_x.
\end{aligned}
$$

Whence, in the linear case, $\omega D = -Q'(0)$. Here we have assumed there is no flux when there are no gradients, and the flux is an odd function of the gradient. Now, we instead impose a constitutive assumption on the dispersive flux that maintains finite flux even when the gradients become large. In particular, we assume that the flux is given by

$$\phi = -\omega D Q \left(\frac{L}{C_0} C_x \right) + VC, \tag{3.23}$$

where L is a given length scale in the problem and Q is a given continuously differentiable function with the "flux" properties

$$|Q(r)| \leq M, \quad Q(0) = 0, \quad Q'(r) > 0, \quad Q'(r) \to 0 \text{ as } |r| \to \infty. \tag{3.24}$$

For example, both $Q(r) = \tanh r$ and $Q(r) = r/\sqrt{1+r^2}$ satisfy the requirements. Under this bounded flux constitutive assumption, the governing mass balance equation becomes

$$C_t = DQ \left(\frac{L}{C_0} C_x \right)_x - vC_x - \frac{\rho_b}{\omega} F(C/C_0)_t.$$

If we rescale using the dimensionless quantities

$$\xi = \frac{x}{L}, \quad \tau = \frac{t}{L/v}, \quad u = \frac{C}{C_0},$$

then the model can be written

$$(u + f(u))_\tau = aQ(u_\xi)_\xi - u_\xi, \tag{3.25}$$

where

$$a \equiv \frac{D}{VC_0}, \quad f(u) \equiv \frac{\rho_b}{\omega C_0} F(u)$$

are dimensionless quantities. Here, a measures the ratio of diffusion to convection (the inverse of the Peclet number) and $f(u)$ is a dimensionless form of the equilibrium isotherm. We assume that the isotherm satisfies the usual conditions

$$f \in C^1(\mathbb{R}^+), \quad f(0) = 0, \quad f'(u) > 0, \quad f''(u) < 0 \text{ if } u > 0. \tag{3.26}$$

For specific examples, we can consider the Freundlich and Langmuir isotherms given by

$$f(u) = \sqrt{u}, \quad f(u) = \frac{u}{1+u},$$

respectively.

3.3.1 Smooth Wave Fronts

We seek solutions of (3.5) of the form $u = U(z)$ where $z = \xi - c\tau$ with

$$U(-\infty) = u_l > 0, \quad u(+\infty) = 0, \tag{3.27}$$

where the wave speed c and the state u_l are to be determined. Under this hypothesis (3.25) becomes

$$-c(U + f(U))' = aQ(U')' - U',$$

where *prime* denotes d/dz. The existence of wave fronts depends intimately on the parameter a. If $a \gg 1$, then diffusion dominates and we expect traveling waves. On the other hand, if $a \ll 1$, then there is not enough diffusion to prevent shocks from forming and it is possible to propagate discontinuous wave fronts. Integrating the last equation once and using the boundary condition at $+\infty$ to evaluate the constant of integration (which is zero), we obtain the first-order equation

$$U(1 - c) - cf(U) = aQ(U'). \tag{3.28}$$

The boundary condition at $z = -\infty$ and the fact that $Q(0) = 0$ forces a relationship between u_l and c, namely,

$$c = c(u_l) = \frac{u_l}{u_l + f(u_l)}. \tag{3.29}$$

Thus, if traveling waves exist, the wave speed satisfies the inequality $0 < c < 1$. Because f is increasing, (3.29) can be solved to determine $u_l = u_l(c)$, i.e., the state at minus infinity in terms of the wave speed.

To analyze the differential equation (3.28) let us define the quantity

$$G(U, c) \equiv a^{-1}(U(1 - c) - cf(U)).$$

Then the differential equation (3.28) is

$$Q(U') = G(U, c),$$

which is an implicit differential equation. The issue is when we can solve for the derivative. We think of $G(U, c)$ versus U as a one-parameter family of curves depending on c, or equivalently u_l. Clearly $G(0, c) = G(u_l, c) = 0$ and $G_{UU}(U, c) = -ca^{-1}f''(U) > 0$. Therefore, the graph of G versus U is concave upward and negative in $0 < U < u_l$. We calculate its minimum. Using $G_U(U, c) = 1 - c - cf'(U) = 0$ we determine that the minimum occurs at u_1, where u_1 satisfies the equation

$$f'(u_1) = \frac{1 - c}{c} = \frac{f(u_l)}{u_l}.$$

The minimum value is

$$M(c) \equiv \min_{0 \leq U \leq u_l} G(U, c) = a^{-1}((1 - c)u_1 - cf(u_1)),$$

and $M(c)$ decreases with c. For some isotherms (e.g., Langmuir) this minimum is bounded below and for others (e.g., Freundlich) it is not. By assumption, Q is bounded below and thus the right-hand side of the differential equation (3.28), which represents the dispersive flux, is bounded. The left-hand side, which represents the advective and inertial transport, as well as the kinetics, is negative. In order to solve the equation for U' we need $G(U, c)$ to lie in the domain of Q^{-1}. Thus, if $-Q_0 = \inf Q$, then there is a threshold value c^* of c, and a corresponding state u^* at $-\infty$ given by

$$M(c^*) = -Q_0 < 0,$$

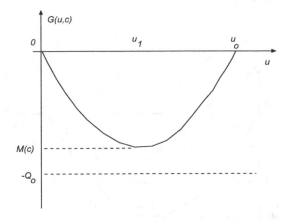

Figure 3.7: Subcritical case: $M(c) \geq \inf Q$.

where the dispersive flux balances the transport and kinetics terms. For $c \leq c^*$ we have $G(U, c) \geq -Q_0$ and $U' = Q^{-1}(G(U, c))$. See figure 3.7; this is the **subcritical case**. The right-hand side of this differential equation is continuously differentiable and there exists a smooth trajectory connecting the states u_l at $z = -\infty$ to zero at $z = +\infty$.

Exercise 58 *In the subcritical case verify that the differential equation $U' = Q^{-1}(G(U, c))$ has stable and unstable equilibrium solutions at $U = 0$ and $U = u_l$, respectively.*

3.3.2 Supercritical Waves

If, for a given value of c, we have $M(c) < \inf Q$, then there exists states u_a and u_b, $0 < u_a < u_b < u_l$, where

$$G(u_a, c) = G(u_b, c) = \inf Q.$$

Note that u_a and u_b depend on c. See figure 3.8; we call this the **supercritical case**. In this case there is a smooth solution in the interval $-\infty < z < 0$ connecting the state u_l to u_b, and a smooth solution in the interval $0 < z < \infty$ connecting the state u_a to the state zero. At $z = 0$ we insert a jump discontinuity of magnitude $[U] = u_b - u_a$. See figure 3.9. It is straightforward to verify that this shock solution is a weak solution to the partial differential equation (3.25); it follows from the fact that the jump condition associated with (3.25) is [e.g., see Logan (1994)]

$$-c[u + f(u)] + [u - aQ(u_\xi)] = 0,$$

where the square brackets notation $[\cdot]$ denotes the jump in the quantity inside (the value behind the jump minus the value ahead). By construction, the

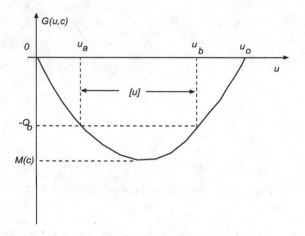

Figure 3.8: Supercritical case: $M(c) < \inf Q$.

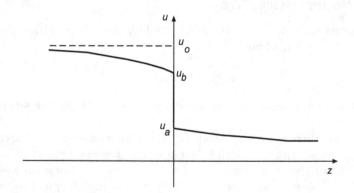

Figure 3.9: Subcritical case: weak solution.

solution $u = U(z)$ satisfies the condition

$$Q(U') = a^{-1}(U(1-c) - cf(U)),$$

so that

$$[Q(U')] = a^{-1}[(U(1-c) - cf(U))],$$

which is, in fact, the required jump condition.

3.3.3 Langmuir Kinetics

Let us illustrate the analysis with some specific isotherms. Consider the equation

$$\left(u + \frac{u}{1+u}\right)_t = a\left(\frac{u_x}{\sqrt{1+u_x^2}}\right)_x - u_x.$$

Here $f(u) = u/(1+u)$ is the Langmuir isotherm, $Q = r/\sqrt{1+r^2}$ is the flux function, and $\inf Q = -1$. The speed of traveling waves is given according to (3.29) by

$$c = \frac{1+u_l}{2+u_l},$$

where u_l is the state at $-\infty$. Clearly, $\frac{1}{2} < c < 1$. Inverting yields

$$u_l = \frac{2c-1}{1-c}.$$

The function $G(U, c)$ is given by

$$G(U, c) = \frac{U}{a}(1 - c - \frac{c}{1+U}),$$

and its minimum occurs at

$$u_1 = \sqrt{\frac{c}{1-c}} - 1.$$

Then the minimum value of G is found to be

$$M(c) = \frac{2\sqrt{c(1-c)} - 1}{a}.$$

Clearly M is a decreasing function on $0.5 < c < 1$ with $M(0.5) = 0$ and $M(1) = -1/a$.

Therefore, if the diffusion constant a satisfies $a \geq 1$, then $M(c) \geq -1 = \inf Q$ for all $c \in (0.5, 1)$, and we obtain smooth traveling wave solutions.

On the other hand, if $a < 1$, then we are in the supercritical case and shock solutions will exist. Fix a. To determine the critical value of c in this case we set $M(c) = -1$ and solve to obtain

$$c = c^* = \frac{1}{2}\left(1 + \sqrt{1 - (1-a)^2}\right).$$

Obviously c^* is an increasing function of a. For example, if $a = 0.1$, then $c^* \approx 0.718$. Thus, in the case $a < 1$ we obtain smooth traveling waves in the range $0.1 < c \leq c^*$ and shock solutions in the case $c^* < c < 1$.

To determine the magnitude of the jump in the supercritical case we set $G(U, c) = -1$, or

$$U(1 - c) - \frac{cU}{1 + U} = -a.$$

Here $a < 1$ is fixed, and c is fixed with $c^* < c < 1$. Solving for u and subtracting the two roots gives the magnitude of the jump as

$$[U] = \frac{\sqrt{(2c - a - 1)^2 - 4a(1 - c)}}{1 - c}.$$

For example, if $a = 0.1$, then $c^* \approx 0.718$. Choose $c = 0.8$. Then the state at $z = -\infty$ is $u_0 = 3$ and the jump in U is computed from the formula to be $[U] = 2.06$.

Observe that these results are consistent with the intuition that in the case of small a there is not enough diffusion to counterbalance the formation of shocks.

Exercise 59 *This exerice asks the reader to verify results in the case of a Freundlich isotherm. We have the equation*

$$(u + \sqrt{u})_t = a(\tanh u_x)_x - u_x,$$

where $f(u) = \sqrt{u}$ and the flux is $Q(u_x) = \tanh u_x$. Clearly $\inf Q = -1$. Show that wave speed is

$$c = \frac{1}{1 + 1/\sqrt{u_l}},$$

where u_l is the state at $-\infty$. Inverting,

$$u_l = \left(\frac{c}{1 - c}\right)^2.$$

Determine that

$$G(U, c) = a^{-1}((1 - c)U - c\sqrt{U}),$$

and its minimum occurs at $u_1 = u_l/4$ and is given by

$$M(c) = -\frac{c^2}{4a(1 - c)}.$$

Set $M(c) = \inf Q$ to find the critical wave speed

$$c^* = 12a + 2\sqrt{a(1 + a)}.$$

In contrast to the Langmuir case discussed previously, M is unbounded below. Thus, for any diffusion constant a there are always both subcritical $(c \leq c^*)$ and supercritical $(c > c^*)$ waves. In the supercritical case the jump in the shock is calculated by subtracting the roots of $G(U, c) = -1$ for fixed a to obtain

$$[U] = \frac{1}{4(1 - c)^2}\left((c + \sqrt{c^2 - 4a(1 - c)})^2 - (c - \sqrt{c^2 - 4a(1 - c)})^2\right).$$

3.3.4 Convergence to Traveling Waves

We can always inquire about the stability, or permanence, of traveling waves. If such waves are not stable, then they will not be observed in nature—small perturbations, ever present, will evolve to destroy the wave form. On the other hand, if the wave is stable, these small perturbations will decay and the wave form will persist. In the next section we will deal with the question of permanence in an analytic manner; often, it can be proved that certain wave fronts are stable. We can also investigate these questions numerically. Below we present numerical, or graphical, evidence suggesting that solutions to the Cauchy problem converge to a traveling wave front in one subcritical case.

Consider initial data $\phi(x)$ to be piecewise continuous with $\phi(x) = u_l$ for $x \leq x_0$ and $\phi(x) = 0$ for $x \geq x_1$. We apply a standard, explicit finite-difference scheme to partial differential equation (3.28). Such schemes are discussed in Appendix A. We take the flux to be $Q(r) = \tanh r$ and we examine (3.28) with Langmuir kinetics.

With $Q(r)$ assigned above and the adsorption isotherm given by $f(u) = u/(1 + u)$ we expand the spatial derivative in (3.28) to obtain

$$(u + \frac{u}{1 + u})_t = a \sec h^2(u_x)u_{xx} - u_x \equiv F(u_x, u_{xx}). \tag{3.30}$$

We proceed using an explicit finite-difference method. Let U_i^k denote the discrete approximation to the exact solution $u(x_i, t_k)$ at the lattice point (x_i, t_k), where $x_i = i\Delta x$ and $t_k = k\Delta t$ are the denote i^{th} spatial and k^{th} time steps, respectively, and Δt and Δx are the time and spatial increments. We take the latter to satisfy a stability condition $\Delta t < \frac{\Delta x^2}{a}$. The spatial derivatives u_{xx} and u_x are approximated with a second-order centered difference and an upstream first-order difference, respectively. Then, replacing the left-hand side of (3.30) with a forward difference gives

$$U_i^{k+1} + \frac{U_i^{k+1}}{1 + U_i^{k+1}} = \Delta t(U_i^k + \frac{U_i^k}{1 + U_i^k}) \tag{3.31}$$

$$+ \Delta t F \left(\frac{U_i^k - U_{i-1}^k}{\Delta x}, \frac{U_{i+1}^k - 2U_i^k + U_{i-1}^k}{\Delta x^2} \right).$$

At each time step k the values of U_i^k, $i = 0, \pm 1, \pm 2, \ldots$ are known, allowing the right-hand side of (3.31) to be calculated as a numerical value, say A_i. This results in the quadratic equation

$$(U_i^{k+1})^2 + (2 - A_i) U_i^{k+1} - A_i = 0, \tag{3.32}$$

which is then solved explicity to obtain the approximation U_i^{k+1} at the subsequent time step.

Figure 3.10 shows how a specific, cubic concentration profile evolves in time for the subcritical case of (3.30) with $a = 0.1$ and the initial upstream state $u_l = 1.5$. It appears that this initial, nonmonotone profile evolves into the

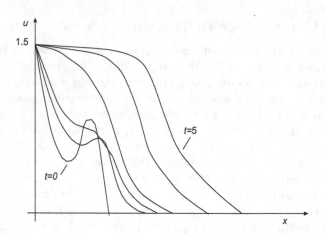

Figure 3.10: Schematic showing how the initial wave profile $u_0(x) = (-1.5x^3 + 4.95x^2 - 4.5x + 1.5)H(2 - x)$ evolves in time.

traveling wave, but further analysis would be required to prove this is the case. The article by Homp and Logan (1999) contains additional examples.

Exercise 60 *In the Freundlich case, assign $Q(r)$ as above and take $f(u) = \sqrt{u}$ to obtain*

$$(1 + \frac{1}{2\sqrt{u}})u_t = a \sec h^2(u_x)u_{xx} - u_x.$$

To avoid computational difficulties that occur on the left-hand side for small u values, solve for u_t before replacing terms with appropriate differences and obtain the difference approximation

$$U_i^{k+1} = U_i^k + \Delta t \frac{2\sqrt{U_i^k}}{1 + 2\sqrt{U_i^k}} F(\frac{U_i^k - U_{i-1}^k}{\Delta x}, \frac{U_{i+1}^k - 2U_i^k + U_{i-1}^k}{\Delta x^2}). \qquad (3.33)$$

Show that numerical computations give results similar to the Langmuir kinetics case, again showing the approach to the traveling wave.

3.4   Stability of Traveling waves

3.4.1   General Ideas

One important issue for any solution to a PDE is it permanence, or its stability. This is particulary true for steady-state solutions to PDEs, of which traveling qualify (they are steady in the moving coordinate system). The question of permanence is the following: If the steady solution is perturbed, or disturbed, by a small amount, does the steady solution persist? That is, do the small perturbations decay away and the system return to its original steady state, or do the perturbations cause the system to change drastically?

For a general illustration and outline of the method, consider the nonlinear scalar parabolic equation

$$u_t = \alpha u_{xx} + g(u, u_x).$$

As we remarked, such equations often admit special traveling wave solutions of the form $u = U(z)$, $z = x - ct$, that satisfy the boundary conditions

$$u \to u_l \quad as \quad z \to -\infty$$

and

$$u \to u_r \quad as \quad z \to +\infty.$$

This wave profile U satisfies the boundary value problem

$$-cU' = \alpha U'' + g(U, U'), \quad U(-\infty) = u_l, \quad U(+\infty) = u_r,$$

or an equivalent dynamical system

$$
\begin{aligned}
U' &= V, \\
V' &= -\alpha^{-1}(cV - g(U, V)).
\end{aligned}
$$

If, at time $t = 0$, we impose a small perturbation on the wave and then inquire about the evolution of that perturbation as $t \to \infty$, can we say anything about its growth or decay properties?

As stated above, we regard the traveling wave as a steady-state solution in the moving coordinate system. If we change variables to the moving frame by taking new independent variables t and $z = x - ct$, then it is easy to check that partial derivatives transform according to

$$\partial_t \to \partial_t - c\partial_z, \quad \partial_x \to \partial_z.$$

Therefore, in the moving frame, the differential equation becomes

$$u_t = \alpha u_{zz} + cu_z + g(u, u_z). \tag{3.34}$$

We say that the traveling wave U is *stable in a norm* $||\cdot||$ if there exists a positive constant δ such that, if $||u(z, 0) - U(z)|| < \delta$, then $||u(z, t) - U(z + h)|| \to 0$ as $t \to \infty$ for some $h \in \mathbb{R}$. In words, this definition says that if the perturbation is sufficiently small initially, then the solution to (3.34) asymptotically approaches some translate of the traveling wave. The closeness is measured in some norm, for example, the $L^2 = L^2(\mathbb{R})$ norm (see Chapter 1) or a *weighted* L^2 norm, and usually the choice of a norm is a crucial issue. By a weighted L^2 norm we mean a norm of the form

$$||f|| = \left(\int_{\mathbb{R}} |f(x)|^2 W(x) dx \right)^{1/2},$$

for some given, positive weight function $W(x)$.

The approach we take in this monograph is based on **linearization**. We assume the perturbations about the traveling wave are small and then we obtain a linearized equation for the perturbations by discarding the presumed small nonlinear terms. Spectral analysis of the linear equation then leads to stability results. To be more precise, let

$$u = U(z) + w(z,t),$$

where $w(z,t)$ is a small perturbation, and $w(z,0)$, the initial perturbation, is given. Then, by substitution, w satisfies the **nonlinear perturbation equation**

$$w_t = \alpha w_{zz} + \alpha U'' + cU' + cw_z + g(U+w, U'+w_z),$$

or, using the equation for U,

$$w_t = \alpha w_{zz} + cw_z + g(U+w, U'+w_z) - g(U, U').$$

Now, expanding the function g in its Taylor series about (U, U'), we have

$$g(U+w, U'+w_z) - g(U, U') = g_u(U,U')w + g_{u_z}(U,U')w_z + \text{ small terms}.$$

Therefore, discarding the assumed small nonlinear terms, we obtain the **linearized perturbation equation**

$$w_t = \alpha w_{zz} - a(z)w_z - b(z)w, \quad w(z,0) = w_0(z),$$

where

$$a(z) \equiv -c - g_u(U, U'), \quad b(z) \equiv -g_u(U, U').$$

Let us write this simply as

$$w_t = -Aw, \ \ t > 0, \ z \in \mathbb{R}; \quad w(z,0) = w_0(z), \ \ z \in \mathbb{R},$$

where A is the differential operator defined by

$$Aw = -\alpha w_{zz} + a(z)w_z + b(z)w.$$

The problem is to determine, for stability, if the perturbation w decays in some norm as $t \to \infty$.

On an infinite interval this can be a difficult problem to resolve. Before discussing it further, let us briefly review the finite-interval case. On a finite, bounded interval I the spectrum consists only of eigenvalues and a modal argument can made. That is, if we assume modal solutions (**Fourier modes**) of the form $w = e^{-\lambda t}y(z)$, then we obtain the eigenvalue problem

$$Ay = \lambda y, \quad z \in I.$$

Here, we are assuming boundary conditions at the endpoints of the interval, for example, the Dirichlet conditions. (See Section 1.4.) If all of the eigenvalues

$\lambda = \lambda_n$ lie in the right half plane with corresponding eigenfunctions $y = y_n(z)$, then all the modal solutions decay as $t \to \infty$. Thus, the solution, which can be written as a superposition of all the Fourier modes as

$$w(z, t) = \sum c_n e^{-\lambda_n t} y_n(z),$$

decays as $t \to \infty$.

On an infinite interval ($z \in \mathbb{R}$) the eigenvalue problem is singular and not so simple to solve. Not only are the coefficients variable, but the problem is usually not self-adjoint. In this case the spectrum of the operator A can be composed of complex numbers other than eigenvalues. A complete discussion of these matters would take us beyond the scope of this monograph and into the theory of unbounded operators on Hilbert spaces. The reader who wants an accessible, modern, applied introduction to these ideas can consult Renardy and Rogers (1993); for a more general treatment specific to parabolic equations see Henry (1981).

Very briefly, let us introduce some of the main issues. Let $A : D(A) \subset L^2(\mathbb{R}) \to L^2(\mathbb{R})$ be a linear operator with domain $D(A)$. The **resolvent** of A is the operator $R_\lambda \equiv (A - \lambda I)^{-1}$, where I is the identity operator. A number λ is called an **eigenvalue** of A if R_λ does not exist, i.e., if the operator $A - \lambda I$ is not one-to-one. This means, of course, that $Ay = \lambda y$, or equivalently $(A - \lambda I)y = 0$, will have nontrivial solutions when λ is an eigenvalue. The **resolvent set** is the set of λ for which R_λ does exist, is bounded, and its domain is dense. The **spectrum** of A is the complement of the resolvent set, which of course includes the eigenvalues, or **point spectrum**. But there may be more to the spectrum than the eigenvalues; the resolvent operator may be unbounded or not densely defined. If R_λ exists and is densely defined, but unbounded, we say λ belongs to the **continuous spectrum**; if R_λ exists and is bounded, but not densely defined, then λ belongs to the **residual spectrum**. Thus, the complex plane is partitioned into four disjoint sets: the resolvent set, the eigenvalues, and the residual and continuous spectrums. The noneigenvalues in the spectrum is sometimes called the **essential spectrum**. Consequently, for example, for λ to get into the continuous spectrum, the problem

$$(A - \lambda I)y = f$$

must be solvable for most f [in a dense subset of $L^2(\mathbb{R})$], but $(A - \lambda I)^{-1}$ is not bounded. One can show that, if λ is in the continuous spectrum, then there exists a sequence of functions y_n of unit norm in $D(A)$ for which $||(A - \lambda I)y_n|| < 1/n$, $n = 1, 2, 3....$. In other words, λ is very close to being an eigenvalue. The continuous spectrum gets its name because it sometimes forms a an interval, or continuum, in the complex plane, unlike eigenvalues, which are discrete.

It is not surprising, therefore, that values of λ in the continuous spectrum must be considered when examining stability questions. It may occur that a element of the continuous spectrum lies in the right half plane and thus leads to a growing Fourier mode. Consequently, to prove stability we must tie down the location of both the eigenvalues and the continuous spectrum. A theorem

of Weil, stated in the next section, permits us to locate the essential spectrum
for certain differential operators. The statement of Weil's theorem will require
the notion of a closed operator, a definition we give presently. An operator
A is **closed** if for every sequence $y_n \in D(A)$ with the properties $y_n \to y$ and
$Ay_n \to f$, we must have $y \in D(A)$ and $Ay = f$. It is a fact that a differential
operator with smooth coefficients can be extended to a closed, densely defined
operator.

Why does stability involve calculating the entire spectrum? It is beyond
the scope of this monograph to go into these details, but we briefly indicate
the answer. When there are only eigenvalues, the solution can be represented
as a Fourier series in the eigenfunctions, as above. When there is continuous
spectrum as well, we might expect the solution to the linearized perturbation
problem to be a superposition of the form

$$w(z,t) = \sum c_n e^{-\lambda_n t} y_n(z) + \int c(\lambda) y(z, \lambda) e^{-\lambda t} d\lambda,$$

where the sum is over the n for which λ_n is an eigenvalue (i.e., the point spectrum), and the integral is over the continuous spectrum and $y(z, \lambda)$ are "improper" eigenfunctions corresponding to points λ in the continuous spectrum.
In either case, positive λ, or λ with positive real part, can lead to nondecaying
modes.

The reader should recall that linear operators on finite-dimensional spaces,
usually represented by matrices, have a spectrum consisting only of eigenvalues.
Therefore the preceding ideas arise only for operators, like differential or integral operators, acting on infinite-dimensional spaces, like $L^2(\mathbb{R})$. Self-adjoint
operators have no residual spectrum.

The linearization technique is a common method for proving stability. If we
can determine the spectrum, then we can determine decay rates for the solution
of the nonlinear problem in a local neighborhood of the wavefront. However,
we must be sure that the problem admits a meaningful linearization. This
can be a difficulty in the case of degenerate equations. In these instances
different techniques have been developed using the behavior of solutions of the
nonlinear equation itself. These methods, based on ideas involving semigroups
of operators, often provide an explicit domain of attraction for the wave front,
which is not the case for linearization techniques. The origin of this technique for
advection-diffusion equations is the paper by Osher and Ralston (1982) [see also
Hilhorst and Hulshof (1991) and Hilhorst and Peletier (1997)]. A systematic
development of the semigroup method can be found in Peletier (1997).

Finally we remark that it is not always a traveling wave that is the attractor.
In many interesting problems solutions approach a similarity solution in long
time. See, for example, van Duijn and de Graaf (1987).

Exercise 61 *We showed in the text that Fisher's equation $u_t = u_{xx} + u(1 - u)$
admits a traveling wave solution $u = U(z)$, $z = x - ct$, when $c > 2$. In this case,
show that the linearized perturbation equation in the zt-frame is given by*

$$w_t - w_{zz} - cw_z = (1 - 2U(z))w.$$

Suppose the perturbations are confined to the interval $(-l, l)$ and the boundary condtions $w(-l, t) = w(l, t) = 0$ hold for all t. If $w = y(z)exp(-\lambda t)$, show that y satisfies the boundary value problem

$$y'' + cy' + (\lambda + 1 - 2U(z))y = 0, \quad y(-l) = y(l) = 0.$$

Demonstrate that the eigenvalues λ are strictly positive, thus showing linearized stability in this idealized case of constricted perturbations.

3.4.2 L^2-Stability of Traveling Waves

In this section we use the linearization technique described above to show that traveling waves in the equilibrium model are stable to small perturbations.

It is clear that the traveling wave solution $u = U(z)$ must satisfy the non-linear differential equation

$$-c(1 + \alpha f'(U))U' = DU'' - U'. \tag{3.35}$$

and the model equation (3.34) in the moving, traveling wave frame, is given by

$$(1 + \alpha f'(u))(u_t - cu_z) = Du_{zz} - u_z, \quad z \in \mathbb{R}. \tag{3.36}$$

Substituting

$$u = U(z) + v(z, t)$$

into (3.36), where v is a small perturbation in the moving frame, we find that the linearized equation governing the small perturbation is

$$(1 + \alpha f'(U))v_t = Dv_{zz} + (c - 1 + \alpha cf'(U))v_z + \alpha cU'f''(U)v. \tag{3.37}$$

Letting

$$v(z, t) = \phi(z)e^{-\lambda t}$$

we obtain the non-self-adjoint, singular eigenvalue problem

$$L\phi \equiv -\phi'' + a(z)\phi' + b(z)\phi = \lambda d(z)\phi, \quad z \in \mathbb{R}, \tag{3.38}$$

where

$$\phi \in L^2(\mathbb{R}),$$

and the coefficients are given by

$$\begin{align}
a(z) &= (1 - c - \alpha cf'(U))/D, \\
b(z) &= -\alpha cU'f''(U)/D, \\
d(z) &= (1 + \alpha f'(U))/D.
\end{align}$$

The goal now is to determine the spectrum of this singular operator; it consists of both point spectrum (eigenvalues) and essential spectrum. In determining the latter, we require the following important theorem [see Henry (1981)]. We state the theorem for vector fuctions and systems of equations, although we shall use it only in the scalar case.

Theorem 62 Let $M(z)$ and $N(z)$, $z \in R$, be bounded, real $n \times n$ matrices with

$$M(z) \to M_\pm, \quad N(x) \to N_\pm, \quad z \to \pm\infty,$$

where M_+, M_-, N_+ and N_- are constant matrices. Let D be a constant, symmetric, positive definite matrix, and let A be a closed, densely defined, vector valued operator on $L^2(R)$ defined by

$$A\phi \equiv -D\phi'' + M(z)\phi' + N(z)\phi.$$

Then the essential spectrum of A lies inside or on the algebraic curves $S_\pm$ defined by

$$S_\pm = \{\lambda \in \mathbb{C} : \det(k^2 D + ikM_\pm + N_\pm - \lambda I) = 0, \quad k \in \mathbb{R}\}.$$

We first transform (3.38) to eliminate the $d(z)$ term on the right-hand side. Let

$$x = \int_0^z \sqrt{d(\zeta)}d\zeta,$$

which is a one-to-one and onto transformation. Then, in terms of the independent variable x, equation (3.38) becomes

$$A\phi \equiv -\phi'' + m(x)\phi' + n(x)\phi = \lambda\phi, \quad x \in \mathbb{R}, \tag{3.39}$$

where $\phi \in L^2(\mathbb{R})$, and where

$$m(x) = \frac{a(z)}{\sqrt{d(z)}} - \frac{d'(z)}{2d(z)^{3/2}}, \quad n(x) = \frac{b(z)}{d(z)}.$$

Here, $z = z(x)$ through the inverse transformation.

Using the $+$ and $-$ subscripts to denote the constant limits of a function at $+\infty$ and $-\infty$, respectively, one can show that $a(z), b(z)$, and $d(z)$ have the properties

$$b_\pm = 0, \quad d'_\pm = 0, \quad 0 < d_- < d_+, \quad a_- > 0, \quad a_+ < 0.$$

Therefore

$$m_\pm = \frac{a_\pm}{\sqrt{d_\pm}}, \quad n_\pm = 0.$$

By Theorem 62 the essential spectrum of A in (3.39) lies on or to the right of the two parabolas

$$S_\pm = \{\lambda \in \mathbb{C} : \lambda = k^2 + ikm_\pm, \quad k \in \mathbb{R}\}.$$

Because both parabolas contain $\lambda = 0$, we are unable to bound the essential spectrum away from zero. The remedy for this is to transform the problem to a weighted L^2 space to get the properties we need.

To this end, let w be a positive, smooth function (which is to be determined) and define the weighted L^2 space $X = L^2_w(\mathbb{R})$ with norm defined by

$$\|\psi\|^2_X \equiv \int_{-\infty}^{\infty} |\psi(x)|^2 w(x)^{-2} dx.$$

Note that, if $\phi = \psi/w$, then $\psi \in X$ if and only if $\phi \in L^2$. We want to restrict the operator A in (3.39) to $\psi \in X$. Clearly, a function $\psi \in X$ is an eigenfunction if and only if there exists λ such that $A\psi = \lambda\psi$ or

$$w^{-1}A(w\phi) = \lambda\phi, \quad \phi \in L^2.$$

Now define the associated operator B by $B \equiv w^{-1}A(w \cdot)$. Then $\psi \in X$ is an eigenfunction of A iff $B\phi = \lambda\phi, \quad \phi \in L^2$. By a straightforward calculation,

$$B\phi = -\phi'' + (m - 2w'/w)\phi' + (n + mw'/w - w''/w)\phi.$$

Therefore, choose w such that $w'/w = m/2$, or

$$w(x) = e^{\int_0^x (m(\zeta)/2)d\zeta}.$$

Then $w''/w = m^2/4 + m'/2$ and

$$B\phi = -\phi'' + q(x)\phi, \quad q(x) \equiv n + m^2/4 - m'/2. \tag{3.40}$$

We also observe that the essential spectrum is preserved in transforming to the weighted space; this follows from the calculation that

$$(A - \lambda I)\psi = f, \quad \psi, f \in X$$

is equivalent to

$$(B - \lambda I)\phi = g \equiv fw^{-1}, \quad \phi, g \in L^2.$$

Now we apply Theorem 62 to the associated operator B in (3.40). The essential spectrum lies to the right of the two parabolas

$$S_{\pm} = \{\lambda \in C : \lambda = k^2 + q_{\pm}, \ k \in \mathbb{R}\}.$$

Therefore, we must calculate $q_{\pm}$. With a little effort we find

$$q_{\pm} = \frac{a_{\pm}^2}{4d_{\pm}} > 0,$$

and therefore *the essential spectrum of B lies strictly in the right half plane.*

Now we consider the location of the point spectrum. We give two different arguments in this example, one based on an energy estimate and one based on a maximum principle, to determine the location of the point spectrum of B.

We first use an energy argument to show that the nonzero eigenvalues of B, if they exist, lie in the right half plane. We consider the eigenvalue problem

$$B\phi = -\phi'' + q(x)\phi = \lambda\phi, \quad \phi \in L^2.$$

Multiplying by ϕ, integrating over the real line, and then integrating the $\phi\phi''$ term by parts, we obtain

$$\lambda \int_{-\infty}^{\infty} \phi^2 dx = \int_{-\infty}^{\infty} (\phi'^2 + q(x)\phi^2) dx.$$

But now let $\psi = U'/w$, where U is the traveling wave and w is the weight function. We note that U' is of one sign (negative), which is essential in the following arguments. Using the fact that $AU' = 0$ (that is, zero is an eigenvalue of A with eigenfunction U'), it is straightforward to show that $\psi''/\psi = q(x)$. Therefore

$$
\begin{aligned}
\lambda \int_{-\infty}^{\infty} \phi^2 dx &= \int_{-\infty}^{\infty} (\phi'^2 + \psi''\phi^2/\psi) dx \\
&= \int_{-\infty}^{\infty} (\phi'^2 - 2\phi\phi'\psi'/\psi + \psi'^2\phi^2/\psi^2) dx \\
&= \int_{-\infty}^{\infty} \psi^2 \left(\frac{d}{dx}\left(\frac{\phi}{\psi}\right) \right)^2 dx.
\end{aligned}
$$

In the second step above we integrated by parts and used the asymptotic properties of w and U' to get the boundary term to vanish. Therefore, if λ, ϕ is an eigenpair ($\lambda \neq 0$), then we must have $\lambda > 0$.

Now we present an alternate proof based on the maximum principle. As before, let $\psi = U'/w$. Then $B\psi = w^{-1}A(w\psi) = w^{-1}A(U') = 0$. Thus $\psi \in L^2$ is an eigenfunction of B with corresponding zero eigenvalue. Now, by way of contradiction, suppose there exists $\phi \in L^2$ such that $B\phi = \lambda\phi$ with $\lambda < 0$. It is well known from the asymptotic behavior of the solution that $\phi \sim \exp(-\sqrt{q_+ - \lambda}x)$ as $x \to +\infty$ and $\phi \sim \exp(\sqrt{q_- - \lambda}x)$ as $x \to -\infty$. Now let

$$r(x) = \frac{\phi(x)}{\psi(x)}.$$

From the asymptotic behavior of ψ it follows that $r \to 0$ as $x \to \pm\infty$. It also follows that $r = r(x)$ satisfies

$$r'' + 2\frac{\psi'}{\psi}r' + \lambda r = 0, \quad x \in \mathbb{R}.$$

Now suppose r has a positive maximum at $x = \xi$. Then $r''(\xi) \leq 0$, $r'(\xi) = 0$, and $r(\xi) > 0$. But $r''(\xi) = -\lambda r(\xi) > 0$, which is a contradiction. Similarly, r cannot have a negative minimum. Consequently, $r \equiv 0$ and so $\phi \equiv 0$. Thus there are no negative eigenvalues.

In summary, we have shown that perturbations in the weighted space X are linearly stable.

Exercise 63 *Verify that $\lambda = 0$ is an eigenvalue of A with corresponding eigenfunction U'.*

Exercise 64 *Consider the constant coefficient operator on $L^2(\mathbb{R})$ given by*

$$A\phi = -\phi'' + 2a\phi' + b\phi.$$

Apply Theorem 62 to isolate the essential spectrum of A. Show that, if A is confined to the weighted space X with norm

$$\|f\|_X = \left(\int_{-\infty}^{+\infty} |f(z)|^2 e^{2\gamma z} dz \right)^{1/2},$$

for some γ, then it induces an operator B on $L^2(\mathbb{R})$ defined by

$$B\phi = e^{\gamma z} A(e^{-\gamma z}\phi),$$

and locate the essential spectrum of B. Determine γ for which the essential spectrum of A is in the right half plane when restricted to the weighted space X. [For example, see Grindrod (1996).]

Exercise 65 *Consider the differential operator*

$$A\phi = -\phi'' - c\phi' + (2U(z) - 1)\phi, \quad \phi \in L^2(\mathbb{R}),$$

which arose in the study of Fisher's equation, where $U(z)$ is the traveling wave solution to Fisher's equation, $c > 2$. Use Theorem 62 to locate the essential spectrum of A by finding the algebraic curves $S_\pm$. Show that in the weighted space with norm

$$\|f\|_X = \left(\int_{-\infty}^{+\infty} |f(z)|^2 e^{cz} dz \right)^{1/2},$$

the essential spectrum gets pushed into the right half plane.

3.5 A Bioremediation Model

A new technology for cleaning underground contamination involves stimulating the indigenous bacteria in the soil to consume the contaminant as a food source and produce nontoxic, neutral products in the process. This restoration technology is called *in situ* bioremediation, and some researchers report a significant decrease in the time required to remediate an aquifer over traditional flushing methods. In this section we present a simple mathematical model of this process. The actual process involves complex chemical, biological, and fluid mechanical interactions, but a model can often shed light on these processes and help us understand some of the interactions. The model we study here was developed by Odencrantz, *et al.* (1993); see also Oya and Valocchi (1997) where additional references can be found.

The idea is this. We consider a porous medium, e.g., soil, of constant porosity and containing an immobile biomass (bacteria) attached to the soil and having an initial, background, constant density M_0 throughout the domain $x \in$

$\mathbb{R}$. In the aqueous solution, moving at constant average velocity v, is a mobile, adsorbing substrate of density $S = S(x,t)$, which represents the contaminant, e.g., a hydrocarbon. Initially it has a uniform distribution throughout the medium. The bioremediation process is to introduce at the left, input boundary (here at $-\infty$) a constant concentration A_0 of nonabsorbing electron acceptors (for example, oxygen), which we regard as a nutrient. The liquid transports the substrate and the nutrient to the solid biomass on the soil where consumption of the substrate by the bacteria takes place. This process can be represented symbolically by the reaction

$$\mathbb{M} + \mathbb{S} + \mathbb{A} \rightarrow \text{neutral products},$$

where $\mathbb{M}$, $\mathbb{S}$, and $\mathbb{A}$ represent the biomass, substrate, and nutrient, respectively. The rate F that this reaction occurs depends on the concentrations $M = M(x,t)$, $S = S(x,t)$, $A = A(x,t)$ of these three substances in a way we shall define subsequently.

The governing equations are the three mass balance equations for the three substances, and they are derived in the same manner as the dispersion–advection–adsorption equations in Section 2.1. We assume that the substrate disperses, advects with the liquid at average velocity v, and adsorbs to the solid matrix. The nutrient disperses and advects, but it does not adsorb. The biomass is immobile, but in the absence of the acceptor or substrate, it decays at a rate proportional to its excess density. Consequently, we have the balance laws

$$\begin{aligned} RS_t &= DS_{xx} - vS_x - F, \\ A_t &= DA_{xx} - vA_x - rF, \\ M_t &= yF - b(M - M_0). \end{aligned}$$

Here, D is the dispersion constant, b is the relative decay rate for the bacteria, y is the yield coefficient (mass of bacteria produced per mass of substrate degraded), and r is the mass of nutrient used per mass of substrate degraded. $R > 1$ is the retardation constant (see Section 2.2) for the adsorption of the substrate. The biogradation rate F is given by the **Monod kinetics** formula

$$F = F(S, A, M) = \frac{qSA}{(K_S + S)(K_A + A)} M. \tag{3.41}$$

Observe that the rate law is Langmuir-like in S and A, and linear in M. The constant q is the maximum specific rate of substrate utilization, and K_S and K_A are the half-maximum rate concentrations of S and A. Note that the substrate and nutrient equations are advection–dispersion equations; the substrate equation has adsorption, the nutrient equation does not. The biomass equation has no transport terms.

Now let us return to the bioremediation scenerio. When the nutrients and substrate reach the biomass, a biologically active zone forms and supports the growth of the microbes. At the same time, the substrate is degraded and the nutrient is consumed. Outside this zone the microbes decrease because of decay.

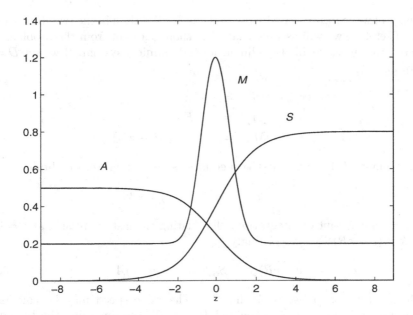

Figure 3.11: Moving biologically active zone involving the biomass, nutrient, and substrate.

It seems reasonable that this zone moves downstream in a pattern-form like the one shown in the schematic in figure 3.11.

This scenario then suggests the possibility of a traveling wave solution with boundary conditions given by

$$
\begin{aligned}
S &= 0, \quad A = A_0 \quad \text{at} \quad -\infty, \\
S &= S_\infty, \quad A = 0 \quad \text{at} \quad +\infty, \\
M &= M_0 \quad \text{at} \quad \pm\infty.
\end{aligned}
\tag{3.42}
$$

The constant A_0 is the nutrient input at the left boundary, and M_0 is the background microbe population. S_∞ is the superimposed substrate density; at the left input boundary, at minus infinity, it has degraded completely. Therefore, let us assume wave forms

$$
S = S(x - ct), \quad A = A(x - ct), \quad M = M(x - ct),
$$

where c is the wave speed. Then we have

$$
\begin{aligned}
-cRS' &= DS'' - vS' - F, \\
-cA' &= DA'' - vS' - rF, \\
-cM' &= yF - b(M - M_0),
\end{aligned}
$$

where differentiation is with respect to the moving coordinate $z = x - ct$. Right away we observe that the system is three-dimensional, involving all three variables. Problems of this type require a more complicated analysis to prove the existence of wave fronts [see Murray and Xin (1998)] than in the two-dimensional case. Therefore we will assume that dispersion is absent from the problem and reduce it to a manageable two-dimensional dynamical system. If we set $D = 0$, we obtain

$$
\begin{align}
(v - cR)S' &= -F, \tag{3.43}\\
(v - c)A' &= -rF, \tag{3.44}\\
-cM' &= yF - b(M - M_0). \tag{3.45}
\end{align}
$$

If we eliminate F from the first two equations and integrate, we obtain

$$(v - cR)S = r^{-1}(v - c)A + k,$$

where k is a constant of integration. Evaluating the last expression at $z = +\infty$ gives $k = (v - cR)S_\infty$. Accordingly,

$$(v - cR)(S - S_\infty) = r^{-1}(v - c)A, \tag{3.46}$$

which is a relation between A and S. The wave speed may be calculated by evaluating (3.46) at $-\infty$, provided $R > 1$. We obtain, using the boundary conditions (3.41),

$$c = v\frac{r^{-1}A_0 + S_\infty}{r^{-1}A_0 + RS_\infty}. \tag{3.47}$$

Note that $0 < c < v$ and $v - cR < 0$. Using (3.46) we are able to eliminate S from the equations (3.43)–(3.45) and obtain a two-dimensional system. In a straightforward manner we obtain

$$
\begin{align}
(v - c)A' &= -rF(S_\infty + \alpha A, A, M), \tag{3.48}\\
cM' &= -yF(S_\infty + \alpha A, A, M) + b(M - M_0), \tag{3.49}
\end{align}
$$

where $\alpha = (v - c)/[r(v - cR)]$. The actual form of F, after substitution, is

$$F = \frac{qMA(S_\infty + \alpha A)}{(k_S + S_\infty + \alpha A)(k_A + A)}.$$

Observe that $S_\infty + \alpha A_0 = 0$. We may now study (3.48)–(3.49) in an AM-phase plane and show there is a unique heteroclinic orbit connecting the equilibrium state (critical point) (A_0, M_0) at $z = -\infty$ to the equilibrium state (critical point) $(0, M_0)$ at $z = +\infty$. See figure 3.12. Because F is nonnegative, it is clear that $A' = 0$ on the vertical nullclines $A = 0, A = A_0, M = 0$, and $A' < 0$ for all A in the strip $0 < A < A_0$. Furthermore, $M' < 0$ for all $0 \leq M \leq M_0$. So, in the first quadrant the horizontal nullcline $M' = 0$ must lie above the line $M = M_0$ and pass through the two critical points. The actual form of the horizontal nullcline is

$$M = M_0\left[1 - \frac{yq}{b}\frac{A}{k_A + A}\frac{S_\infty + \alpha A}{k_S + S_\infty + \alpha A}\right]^{-1}.$$

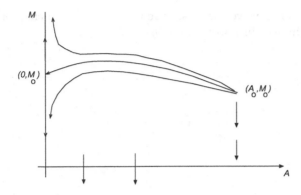

Figure 3.12: Phase portrait for (3.48)–(3.49).

There are two cases: The nullcline is a smooth, continuous curve connecting the critical points, or the nullcline graph of M versus A blows up with one or two vertical asymptotes (the denominator vanishes at none, one, or two points, at most). The first case is shown in the phase plane in figure 3.11; the critical point (A_0, M_0) is an unstable node and $(0, M_0)$ is a saddle point. Hence, the traveling wave exists as the node–saddle orbit connecting the points. Note that M increases in the biological zone and the wavefronts have the qualitative behavior shown in figure 3.12. In the case the vertical nullcline has asymptotes, then one can draw the same conclusions. We leave this as an exercise. Thus, the only condition required for wave front solutions in the zero-dispersion case is that the retardation constant R must satisfy $R > 1$; that is, the substrate must adsorb. This has the effect of causing the substrate to advect at the retardation speed v/R, which is slower than the advection speed v of the nutrient. This difference in transport velocity leads to a spatial zone where the substrate and nutrient are simultaneously present in large enough concentrations to induce reaction and subsequent microbial growth.

3.6　Reference Notes

Wave fronts play an important role in many areas of mathematics and applied science, and the literature on traveling waves is vast. For a general treatment see Grindrod (1996), Logan (1994), and Volpert and Volpert (1994). For biological examples as well as a general discussion, see Murray (1993) and Britton (1986). In combustion, the paper by Diaz and Stakgold (1995) is relevant to the discussions here. For applications in hydrogeological settings we give several citations in the references below. The list is incomplete, but it should set a stepping stone to a complete literature search. Of particular interest are the papers by van Duijn and Knabner (1991, 1992a,b). Peletier (1997) contains many additional references. A nonlocal example is given in Logan (2001). Further

examples of traveling waves are also given in the next two chapters, which deal
with porosity changes in porous domains.

Chapter 4

Filtration Models

Intuitively, filtration occurs in a porous medium when the transported particles are actually sieved by the solid porous fabric, thereby decreasing the porosity and possibly clogging the medium (colmatage). This clearly can, and will, occur when the average size of the pores is smaller than that of the particles in solution, as in the case of large molecules, bacteria or colloids, migrating through clayey soils. But, in hydrogeology and in chemical engineering, filtration also refers to the mechanisms of retention when smaller particles form sediment in domains with large porosity. For example, deep bed filtration is the practice of removing suspended particles from a fluid stream by passing the suspension through beds composed of granular material. As the suspension flows through the granular bed, some of the particulants come in contact with the filter grains and become deposited onto the grains and onto particles already accreted. These changes affect the flow of the suspension through the porous medium and the subsequent deposition. In this chapter we introduce some models of filtration in the latter case. The difference between the adsorption models discussed in Chapters 1 through 3 and the filtration models discussed below is that filtered particles have volume, and when these particles are accreted by the soil matrix, the porosity of the porous domain decreases. Thus, the porosity is not constant. In the adsorption problems studied earlier the porosity remained constant because the volume of the accreted particles, e.g., ions, was considered negligible.

Clearly, filtration processes play an important role in environmental studies. Colloids (large dissolved organic molecules like humic substances, microorganisms, and other mineral matter) may be deposited in aquifers through surface contamination or leaking underground storage facilities. There is experimental evidence that certain bacteria and viruses, and even asbestos fibers, can migrate over long distances through the subsurface. Colloids have also been implicated in the unexpected movement of radioactive elements that were normally believed to be relatively immobile. In this case, the elements attach themselves to colloidal particles that, because of their size, seek out the bigger, rapidly conducting pathways in the medium. Many of these problems and effects are not well understood, and they are a rich source of mathematical models. Another case of interest is the filtration of rain water into agricultural fields; frequently, filter cakes form at the surface from small particle suspensions and prevent water seepage into deeper soil layers.

There are also numerous industrial applications of filtration models. For example, in the manufacturing of composite materials a liquid resin is forced into a solid, porous skeleton and, upon curing, forms a composite. This process usually complicated by the dependence of the process upon temperature changes.

4.1 Models for Colloid Transport

4.1.1 The Herzig–Leclerc–LeGoff Model

We begin by introducing a model of colloid transport with retention kinetics first examined by Herzig, Leclerc, and LeGoff (1970). This model is also discussed by de Marsily (1986). We refer to the model as the **HLL model.** Let ω_0 be the free porosity, i.e., the porosity when no particles are present, of a one-dimensional porous domain of cross-sectional area A. Because transported particles are assumed to have volume, we measure the concentration $C = C(x,t)$ of the transported colloid in *volume* of colloid per unit volume of the liquid–colloid suspension. We assume that the density of the colloid particles is the same as that of the carrying liquid. Furthermore, let $\sigma = \sigma(x,t)$ denote the volume of retained, or immobile, colloids per unit volume of porous medium. Let V be the Darcy velocity of the flow, which is the volume of suspension per unit area per unit time, and let $\phi = \phi(x,t)$ denote the volumetric flux of colloids, measured in volume of colloids per unit area per unit time. Here the Darcy velocity is not assumed to be constant. Indeed, if the medium becomes clogged, the flow may terminate. If $\omega = \omega(x,t)$ is the (variable) porosity of the medium, then we may write the **volume balance equation** in an arbitrary section $a \le x \le b$ of the medium as

$$\frac{d}{dt} \int_a^b C\omega A dx = A(\phi(a,t) - \phi(b,t)) - \int_a^b \sigma_t A dx.$$

This simply states that the rate of change of the total volume of colloids in the section must equal the rate that they flow in, minus the rate they flow out, minus the rate the colloids are accreted onto the solid matrix. In the standard way (see Section 2.1) we obtain the volume balance equation in local form as

$$(\omega C + \sigma)_t = -\phi_x.$$

The first constitutive assumption is the form of the porosity ω. We take

$$\omega = \omega_0 - \beta\sigma,$$

where β is a nonnegative constant called the **compaction factor**. Thus, the volume σ of retained colloids occupies a volume $\beta\sigma$ of the pore space. Next, we assume that the rate of retention is proportional to the concentration of colloids in solution and to the velocity of the flow, and it depends perhaps nonlinearly on the volume concentration of colloids already retained; in other words, we impose the kinetics law

$$\sigma_t = \lambda VCF(\sigma/\omega_0),$$

where the positive constant λ is called the **filter coefficient**. The function F is a nonnegative, nonincreasing function of σ/ω_0. In this model decolmatage is negligible. Finally, we assume the volumetric flux of C contains an advective term and a dispersion term. Specifically,

$$\phi = VC - \omega DC_x.$$

Substituting this constitutive relation back into the volume balance equation and collecting the constitutive equations give the **Herzig–Leclerc–LeGoff** model for colloid filtration

$$
\begin{align}
(\omega C + \sigma)_t &= (\omega DC_x)_x - (VC)_x, & (4.1)\\
\omega &= \omega_0 - \beta\sigma, & (4.2)\\
\sigma_t &= \lambda VCF(\sigma/\omega_0). & (4.3)
\end{align}
$$

The filter coefficient λ, which has dimesions of time^{-1}, could depend in a complicated manner on the diameters of the grains and particles, the viscosity of the fluid, the porosity, and other quantities affecting the efficiency of the fabric as a filter. Generally, one would expect a strong dependence on the porosity. Hydrogeologists have fit various functional forms of λ to collected data, and the reader can consult Herzig *et al.* (1970), Rajagopalan and Tien (1979), or de Marsily (1986, p. 275) for references to experimental work.

In general, the Darcy velocity V may depend on time t, but not on the depth x. To see why this is true we write down a *volume balance* law for the suspension, consisting of the particulants and the liquid. For an arbitrary section $a \leq x \leq b$ we have

$$\frac{d}{dt}\int_a^b \omega A dx = A(V(a,t) - V(b,t)) - \int_a^b (\beta\sigma)_t A dx.$$

Note that there is no net dispersion of the suspension; the only flux into the region is advective flux. But $\omega = \omega_0 - \beta\sigma$, and so the balance law becomes $V(a,t) - V(b,t) = 0$, which means

$$V_x = 0 \quad \text{or} \quad V = V(t).$$

Observe that our balance equations are volume balances rather than mass balances; thus, we will not have the usual continuity equation $\omega_t + V_x = 0$. Therefore, there is an additional mechanism besides flow rate that changes the volume, namely, the rate that particles are accreted on the fabric.

So far, this model ignores the driving force for the flow. In the present discussion, we assume the Darcy velocity V is constant. This means, of course, that a pressure gradient and permeability variation are assumed to maintain this constancy. For example, as a medium clogs, the permeability will change and it will take a greater and greater pressure gradient over the length of the domain to keep the velocity constant. The HLL model ignores these influences. One would not expect the HLL model to be valid in situations near total clogging. In a more accurate model, Darcy's law, which relates pressure gradients to the Darcy velocity, comes into play.

4.1.2   Small Dispersion

A common strategy in studying mathematical models is to investigate the behavior of a model in a special limit.[1] If the particles are large, then one can argue that the dispersion term in (4.1) can be neglected, and thus the mass balance equation becomes

$$(\omega C + \sigma)_t = -(VC)_x.$$

In addition, we assume

$$\omega = \omega_0, \quad F(\sigma) = 1.$$

The assumption $F(\sigma) = 1$ is called the **deep bed filtration hypothesis**. Then the HLL model becomes

$$(\omega C + \sigma)_t = -VC_x \qquad (4.4)$$
$$\sigma_t = \lambda VC, \qquad (4.5)$$

which is a simple hyperbolic system. Eliminating σ from the first equation gives

$$C_t + vC_x = -\lambda vC, \qquad (4.6)$$

which is an advection equation with a source (see Chapter 2), and $v = V/\omega$ is the constant, average velocity. We can easily solve this equation for $t > 0$ on the semi-infinite domain $x > 0$ with initial and boundary data

$$C(x,0) = 0, \quad x > 0; \quad C(0,t) = g(t), \quad t > 0. \qquad (4.7)$$

Here, to obtain a continuous solution, we take $g(0) = 0$. The equation propagates signals at constant speed v. Ahead of the leading signal, i.e., for $x > vt$, we clearly have $C(x,t) = \sigma(x,t) = 0$ because the initial concentration is zero. To determine the solution behind the leading signal, that is, in the domain $0 < x < vt$, we transform to characteristic coordinates $\xi = x - vt$, $\tau = t$. Then equation (4.6) becomes

$$C_\tau = -\lambda vC.$$

This integrates immediately to

$$C = \psi(\xi)e^{-\lambda v\tau},$$

---

[1]When making approximations by deleting terms in equations, it is generally a good idea to introduce dimensionless quantities and scale the problem. We shall, however, postpone scaling for the present calculation.

or

$$C(x,t) = \psi(x - vt)e^{-\lambda vt},$$

where ψ is an arbitrary function. Applying the boundary condition gives

$$C(0,t) = g(t) = \psi(-vt)e^{-\lambda vt},$$

which implies

$$\psi(t) = g(t/v)e^{-\lambda t}.$$

Then the solution is given by

$$C(x,t) = g(t - x/v)e^{-\lambda x}, \quad 0 < x < vt.$$

To determine σ in $0 < x < vt$ we integrate the equation

$$\sigma_t = \lambda V C = \lambda V g(t - x/v)e^{-\lambda x}$$

with respect to t from $t = x/v$ (the leading signal) to t to obtain

$$\int_{x/v}^{t} \sigma_t \, dt = \lambda V e^{-\lambda x} \int_{x/v}^{t} g(s - x/v) \, ds.$$

Because $\sigma(x, x/v) = 0$ (at the front), we have

$$\sigma(x,t) = \lambda V e^{-\lambda x} \int_{x/v}^{t} g(s - x/v) \, ds, \quad 0 < x < vt.$$

Therefore we have obtained the solution to the simplified HLL model (4.4)–(4.5) on the domain $x, t > 0$, subject to the initial and boundary data given in (4.7). In the special case that the boundary concentration is constant, that is, $C(0,t) = g(t) = C_0$, the solution behind the leading wave is

$$C(x,t) = C_0 e^{-\lambda x}, \quad \sigma(x,t) = \lambda V C_0 e^{-\lambda x}(t - x/v), \quad 0 < x < vt. \tag{4.8}$$

To summarize, at time $t = 0$ there is no concentration of suspended colloids or attached colloids in the medium. For $t > 0$ we impose a constant colloid concentration on the inlet boundary $x = 0$. This causes a wave front, or leading signal, to propagate along the spacetime path $x = vt$. Ahead of that signal the concentration of suspended and attached colloids remains zero. Behind the front, at any fixed time t, there is exponential decay of the colloid concentration and the retained concentration from their values on the inlet boundary to their values along the front. There is a jump in the suspended concentration at the front. Along the inlet boundary the retained concentration increases linearly with time, which is unphysical. Thus, the model may be valid only for early times.

4.1.3   Variable Rate Law

Another special case of the HLL model on the domain $x > 0$ moves away from a deep bed filtration hypothesis and assumes a kinetic factor of the form

$$F(\sigma/\omega_0) = 1 - \beta\sigma,$$

where ω_0 is a constant porosity. This linear factor is just one such factor that can be considered; various powers of this factor also lead to equations that can be solved analytically [for example, $F = 1 - \beta^2\sigma^2$, $F = \sqrt{1 - \beta\sigma}$, $F = (1 - \beta\sigma)^{3/2}$]. Introducing the scaled concentration variable $s = \sigma/\omega_0$, and again assuming small dispersion $(D = 0)$ and $\omega = \omega_0$, the HLL model becomes

$$
\begin{aligned}
C_t + s_t &= -vC_x, \\
s_t &= \lambda v C(1 - \beta\omega_0 s),
\end{aligned}
$$

where v is the average velocity. The initial and boundary conditions are

$$C(x,0) = s(x,0) = 0, \quad C(0,t) = c_0.$$

To solve the problem we transform to characteristic coordinates, $\tau = t - x/v$, $\xi = x$, which turns the system of equations into

$$s_\tau = -vC_\xi, \tag{4.9}$$

$$s_\tau = -\lambda v C(1 - \omega_0\beta s). \tag{4.10}$$

In the characteristic coordinate system $\tau = 0$ corresponds to the leading signal, and the initial condition is transferred to the line $\tau = -\xi/v$. Thus, time is measured starting from the arrival of the leading signal. Ahead of the leading signal we have the zero solution. The boundary condition at $\xi = 0$ is

$$C(0,\tau) = C_0.$$

Across the leading signal we assume that s is continuous, thus

$$s(\xi,0) = 0.$$

Separating variables in equation (4.10) and integrating from 0 to τ gives

$$s(\xi,\tau) = \frac{1}{\omega_0\beta}(1 - \exp(-\omega_0\beta \int_0^\tau C(\xi,y)dy)). \tag{4.11}$$

Therefore, from equation (4.9) we get

$$s_\tau = \exp(-\omega_0\beta \int_0^\tau c(\xi,y)dy)C = -vC_\xi. \tag{4.12}$$

Evaluating at $\tau = 0$ gives $C(\xi,0) = -vC_\xi(\xi,0)$, which is an ordinary differential equation for C along the front. Easily,

$$C(\xi,0) = C_0 e^{-\xi/v}.$$

Now, from (4.12), we have

$$(\ln C)_\xi = -\frac{1}{v} \exp(-\omega_0 \beta \int_0^\tau C(\xi, y) dy).$$

Differentiating with respect to τ gives

$$(\ln C)_{\xi\tau} = -\omega_0 \beta C_\xi.$$

Then, integrating on ξ gives

$$\frac{C_\tau}{C} = -\omega_0 \beta C + \Psi(\tau),$$

where Ψ is an arbitrary function. Evaluating this expression at $\xi = 0$ forces $\omega_0 \beta C = \Psi(\tau)$. Hence,

$$\frac{C_\tau}{C} = \omega_0 \beta (C_0 - C).$$

This equation can be integrated by separating variables. Using a partial fraction expansion to compute the integrals and then simplifying gives

$$\frac{1}{C_0} \ln(\frac{C}{C_0 - C}) = \omega_0 \beta \tau + \Phi(\xi).$$

The arbitrary function Φ can be evaluated by setting $\tau = 0$. We obtain

$$\Phi(\xi) = \frac{1}{C_0} \ln(\frac{e^{-\xi/v}}{1 - e^{-\xi/v}}).$$

Finally, after some algebra, we find that the concentration of colloids in suspension is

$$\frac{C}{C_0} = \frac{e^{C_0 \omega_0 \beta \tau}}{-1 + e^{\xi/v} + e^{C_0 \omega_0 \beta \tau}}.$$

The retained colloid concentration is then computed from (4.11). It is

$$s = \frac{1}{\omega_0 \beta}(1 - \frac{e^{\xi/v}}{-1 + e^{\xi/v} + e^{C_0 \omega_0 \beta \tau}}).$$

Here, recall that $\tau = t - x/v$, $\xi = x$.

We can interpret the solution as follows. A time snapshot of the concentration C shows a decreasing profile from the constant value C_0 at the inlet boundary to the value $C_0 e^{-t}$ along the front $t = -x/v$. There is a jump in the concentration across the front. The concentration of retained colloids along the inlet boundary $x = 0$ is given by

$$s(0, t) = \frac{1}{\omega_0 \beta}(1 - e^{-C_0 \omega_0 \beta t}),$$

and therefore it increases to a limiting value of $\omega_0 \beta$. Observe that this model is more realistic than the deep bed filtration model where the boundary concentration increases without bound. For fixed t, the retained concentration decreases from it boundary value to its zero value at the front.

4.1.4   Classical Filtration Theory

We end this section with a brief discussion of classical filtration theory. Much of the classical theory is an experimental science that involves semi-empirical, algebraic formulas containing several parameters that are chosen to fit experimental data. An introduction to this empirical approach can be found in Chapter 15 of Logan (1999). The starting place for this theory is the transport equation

$$C_t = DC_{xx} - vC_x + \widetilde{R}, \quad x > 0,$$

where C is the volume concentration of particles (in volume per unit volume of liquid), D is the constant dispersion coefficient, v is the average velocity, and $\widetilde{R}$ is the rate of attachment or detachment of particles to the porous fabric. If the particles are single-sized and n_0 is the constant number of particles per unit volume of particulant, then $N = n_0 C$ is the number concentration of particles, measured in number of particles per unit volume of liquid. Then we can write the transport equation in terms of number concentration as

$$N_t = DN_{xx} - vN_x + R,$$

where $R = \widetilde{R} n_0$. Here the porosity is assumed to be constant, and we should expect this model to be valid in only special flow regimes. If we assume that mass transport controls attachment, rather than chemistry, then we might make the constitutive assumption $R = -ka(N - N_0)$, where k is a mass transport coefficient, a is the surface area per unit volume, and N_0 is a constant surface concentration. If the surfaces are nearly clean, the so-called "clean-bed" assumption, then N is much larger than N_0 and the transport equation becomes

$$N_t = DN_{xx} - vN_x - kaN.$$

Now come some strong assumptions. Assuming steady conditions and neglecting dispersion we obtain the simple equation

$$vN_x = -kaN,$$

which has solution

$$N(x) = N_i e^{-kax/v},$$

where N_i is an inlet concentration at $x = 0$. Therefore, the concentration of particles decreases expontially through the bed $x > 0$. We emphasize that such models may be valid only for limited time ranges when the assumptions hold.

A semiempirical filtration formula in a porous domain composed of packed spheres has been developed by Rajagopalan and Tien (1976, 1979), who compute the decay factor λ (the filter coefficient) in the exponent in terms of filter properties rather than transport coefficients k, a, and v. Their model provides good agreement with experimental data and is often used to predict particle removal in soil column and subsurface groundwater experiments.   The Ragagopolan–Tien equation is

$$N(x) = N_i e^{-\lambda x}, \quad \lambda = \frac{3}{2d}(1 - \omega_0)^{1/3} \alpha \eta,$$

where d is the diameter of the spheres in the fabric, η is the efficiency of the filter, and α is the sticking coefficient. Further semiempirical formulas have been developed for the filter efficiency in terms of particle diffusion, sedimentation, and interception rate. See Logan (1999). As an exercise, one should compare the Ragagopolan–Tien equation with the solution (4.8) of the deep bed filtration model.

4.2   An Attachment–Detachment Model

4.2.1   The Model

Now we examine an attachment-detachment model that allows either a positive or negative particle accretion rate. The mass balance equation is derived in the same way as in the HLL model. With a view of later introducing dimensionless variables, we use ξ and τ for space and time variables. We take the compaction factor $\beta = 1$. Thus, we have

$$(\omega C + \sigma)_\tau = (\omega D C_\xi)_\xi - (VC)_\xi,$$
$$\omega = \omega_0 - \sigma.$$

Here $V = V(\tau)$. As before, we ignore diffusion, but now we include dispersion. A common assumption is $\omega D = \alpha V$, where is the dispersivity (see Chapter 2).

As in the HLL model we assume that the accretion rate is proportional to the Darcy velocity, to the concentration of particles in suspension, and to the concentration of retained particles in a manner to be specified later. There is also a mechanism for which accreted particles can be returned to the suspension at a rate proportional to the Darcy velocity and to the concentration of accreted particles. Thus, we impose the constitutive law

$$\sigma_\tau = \lambda VCF(\sigma/\omega_0) - kV\sigma,$$

where F is a smooth, nonnegative dimensionless function with $F(0) = 1$, and $k > 0$. The rate function F is not necessarily decreasing, as in the HLL model. It could be increasing whenever the accretion rate is a decreasing function of pore size, and one might expect this to be true when the attachment mechanism includes mechanical entrapment.

The Darcy velocity can be eliminated from this one-dimensional model. The idea is to replace time τ by a "temporal" coordinate $\bar{t}$ that measures the total flow through the system (actually, $\bar{t}$ will have length units). We define

$$\bar{t} = \int_0^\tau V(y)dy.$$

It easily follows from the chain rule for derivatives that

$$\frac{\partial}{\partial \tau} = V(\tau)\frac{\partial}{\partial \bar{t}}.$$

Therefore, the model equations become

$$(\omega C + \sigma)_{\bar{t}} = \alpha C_{\xi\xi} - C_\xi,$$
$$\omega = \omega_0 - \sigma,$$
$$\sigma_{\bar{t}} = \lambda C F(\sigma/\omega_0) - k\sigma.$$

The variable velocity does not appear.

This problem can be reformulated in dimensionless variables by defining the scaled quantities

$$x = \lambda\xi, \quad t = \frac{\bar{t}}{\omega_0/\lambda C_0}, \quad \omega = \frac{\omega}{\omega_0}, \quad s = \frac{\sigma}{\omega_0}, \quad u = \frac{C}{C_0},$$

where C_0 is a reference concentration. Then the dimensionless model is

$$((1-s)u + s)_t = a u_{xx} - u_x, \qquad (4.13)$$
$$s_t = F(s)u - bs, \qquad (4.14)$$

where

$$a = \alpha\lambda, \quad b = \frac{k\omega_0}{\lambda}.$$

For the present problem we have scaled C by $C_0 = 1$.

4.2.2 Traveling Waves

Discussions from preceding chapters lead us to think about examining the existence of wave front solutions. Thus, we seek solutions of the form

$$u = u(z), \quad s = s(z), \quad z = x - ct, \qquad (4.15)$$

where c is the unknown wave speed. Both u and s should have constant limits at $z = \pm\infty$. It should cause no confusion in using the same symbols u and s for the wave forms that we used in the model. Observe that these waves are not traveling waves in the normal sense, as t represents the total flow rather than clock time. In laboratory coordinates ξ, τ the wave form will have a fixed shape, but it will have a variable speed. In fact, the concentration in suspension will have the form

$$C = C\left(\xi - \frac{c}{\omega_0}\int_0^\tau V(y)dy\right),$$

and similarly for s. Note that the wave speed c is dimensionless. In the constant Darcy velocity case this reduces to the usual traveling wave form with constant wave speed.

We can identify traveling waves for a variety of physical situations, but we must first determine the possible equilibrium states for the system (4.13)–(4.14). The mass balance equation (4.13) is clearly satisfied for any constant state (u, s), but the rate law (4.14) is satisfied only for a constant state with the property

$$\frac{F(s)}{s} = \frac{b}{u}. \qquad (4.16)$$

Physically, it seems reasonable that this relation should define a function $s = G(u)$ because each possible steady value of the colloid concentration in the mobile phase should correspond to a unique, steady level of attachment. This assumption leads to the mathematical requirement that $F(s)/s$ be monotone decreasing, or

$$F'(s) < \frac{F(s)}{s}. \tag{4.17}$$

Geometrically, this means that the graph of $F(s)$ at any point s cannot be as steep as the chord connecting $(0,0)$ to $(s, F(s))$. All decreasing functions satisfy this requirement, but some increasing functions do as well.

The comparative magnitudes of $F(1)$ and b also affect the set of possible equilibrium states $(\overline{u}, \overline{s})$. If $b > F(1)$, then a unique equilibrium state for all $0 \leq \overline{u} < 1$, and $\overline{s}$ is bounded above by some value smaller than one. In this case there are no equilibruim states corresponding to complete clogging. If $b < F(1)$, then equilibrium states exist only for $\overline{u} < b/F(1)$.

We shall restrict the problem of finding wave front solutions to the case when one of the equilibrium states is $(0, 0)$, corresponding to a lack of particulants. If the zero state is $+\infty$, then we classify the wave as a **contamination wave** (C-wave), and if the zero state is at $-\infty$, then we classify the wave as a **remediation wave** (R-wave). The problem, therefore, is to identify possible end-states $(\overline{u}, \overline{s})$ such that a C-wave or an R-wave connects $(0, 0)$ to $(\overline{u}, \overline{s})$.

Substituting the wave forms (4.15) into (4.13)–(4.14) and then integrating the first equation gives

$$-c((1 - s)u + s)' = au' - u,$$
$$-cs' = F(s)u - bs.$$

The constant of integration was found to be zero by evaluating at the $(0, 0)$ equilibrium state. The other equilibrium state $(\overline{u}, \overline{s})$ must also satisfy the first equation above, and this yields an expression for the wave speed. We get

$$c = \frac{\overline{u}}{(1 - \overline{s})\overline{u} + \overline{s}}.$$

Eliminating $\overline{u}$ from this expression, using (4.16), yields

$$c = \frac{b}{b + g(\overline{s})}, \tag{4.18}$$

where the function g is defined by

$$g(s) = F(s) - bs. \tag{4.19}$$

Because $g(\overline{s}) = (1 - \overline{u})F(\overline{s}) \geq 0$, we have

$$0 < c < 1.$$

In the subsequent discussion we find it convenient to use the function g in place of F.

In summary, therefore, we have the following problem: *Given $a, b > 0$ and a continuously differentiable function $g : [0, 1) \rightarrow \mathbb{R}$ with the properties $g(0) = 1$ and $g'(s) < g(s)/s$, find points $(\overline{u}, \overline{s})$ such that*

$$0 < \overline{s} < 1, \quad g(\overline{s}) > 0, \quad \overline{u} = \frac{b\overline{s}}{b\overline{s} + g(\overline{s})}, \tag{4.20}$$

and show there exists a unique, smooth solution to the system

$$abu' = c[ug(\overline{s}) - bs(1 - u)], \tag{4.21}$$
$$cs' = bs(1 - u) - ug(s), \tag{4.22}$$

with boundary conditions

$$u(-\infty) = u_-, \quad s(-\infty) = s_-, \quad u(+\infty) = u_+, \quad s(+\infty) = s_+,$$

where one of $(u_\pm, s_\pm)$ is $(0, 0)$ and the other is $(\overline{u}, \overline{s})$ with c given by (4.17).

4.2.3   Contamination and Remediation Waves

The existence of traveling wave fronts follows now from the standard phase-plane analysis introduced in Chapter 3. We seek, in the us–phase plane, a heteroclinic orbit connecting two critical points, or equilibrium states. To carry out this program we first classify the critical points by calculating the Jacobian matrix $J = (j_{ik}) = (\frac{\partial(u', s')}{\partial(u, s)})$. We have

$$J(u, s) = \begin{pmatrix} c(g(\overline{s}) + bs)/ab & -c(1 - u)/a \\ -(g(s) + bs)/c & b[1 - u - ug'(s)/b]/c \end{pmatrix}.$$

We have the following well-known facts, which are easily deduced from the expressions

$$\lambda = \frac{1}{2}(\text{tr} J \pm \sqrt{(\text{tr} J)^2 + 4 \det J}) = \frac{1}{2}(j_{11} + j_{22} \pm \sqrt{(j_{11} - j_{22})^2 + 4j_{12}j_{21}})$$

for the two eigenvalues of J. Here, tr $J = j_{11} + j_{22}$ is the trace of J and $\det J$ is the determinant.

Under the assumptions $j_{11} > 0$, $j_{12} < 0$, and $j_{21} < 0$, we have the following facts:

1. The eigenvalues of J are real and can be labeled such that $\lambda_2 < j_{11} < \lambda_1$.

2. $\lambda_2 < 0$ if, and only if, $\det J < 0$.

3. The eigenvectors corresponding to λ_2 lie in the first and third quadrants, while those corresponding to λ_1 lie in the second and fourth quadrants.

Application of these facts yield a qualitative description of the trajectories near the critical points of the system. We summarize the results in two lemmas. The proof of the first follows immediately from the facts (1)–(3).

Lemma 66 *A critical point* (u^*, s^*) *is a saddle point of the system (4.21)–(4.22) if, and only if,* $\det J(u^*, s^*) < 0$; *otherwise, it is an unstable node. In particular,* $(0,0)$ *is a saddle point if, and only if,* $g(\bar{s}) < 1$, *while any other critical point* (u^*, s^*) *is a saddle point if, and only if,* $g'(s^*) > 0$. *If a critical point* (u^*, s^*) *is a saddle point, then the one-dimensional stable manifold (separatrix) corresponding to the negative eigenvalue enters the point* (u^*, s^*) *with a positive slope.*

Lemma 67 *Let* (u^*, s^*) *be a critical point of the system (4.21)–(4.22) with* $0 < u^* < 1, 0 < s^* < 1$, *and let* Q *be the rectangle* $Q = \{0 < u < u^*, 0 < s < s^*\}$. *Then the boundary of* Q *consists entirely of egress points (where the vector field exits the region* Q). *Furthermore, if there is another critical point* $(\tilde{u}, \tilde{s})$ *with* $s^* < \tilde{s} < 1$ *and* $u^* < \tilde{u} < 1$, *then the boundary of the rectangle* $R = \{u^* < u < \tilde{u}, s^* < s < \tilde{s}\}$ *consists entirely of egress points.*

The proof of Lemma 67 involves calculating the signs of the derivatives u' and s' in (4.21)–(4.22) along the four sides of each of the rectangles Q and R in the phase plane. (The reader should sketch these rectangles.) On the lower boundary of Q we have $s = 0$ and so $cs' = -u < 0$. On the line $s = s^*$ we have, after some algebra, $cs' = bs^*(1 - u/u^*)$. Thus, s' is positive on the upper boundary of Q and negative on the lower boundary of R. On the upper boundary of R we have $cs' = b\tilde{s}(1 - u/\tilde{u}) > 0$. Similar calculations yield $au' = -cs < 0$ on the left boundary of Q, $au' = c(1 - u^*)(s^* - s)$ on the line $u = u^*$, and $au' = c(1 - \tilde{u})(\tilde{s} - s) > 0$ on the right boundary of R.

Now, a C-wave is a solution of (4.21)–(4.22) satisfying the boundary conditions

$$u(-\infty) = \bar{u}, \; s(-\infty) = \bar{s}, \; u(+\infty) = s(+\infty) = 0$$

for some $(\bar{u}, \bar{s})$ satisfying (4.20). The phase plane trajectory must connect $(\bar{u}, \bar{s})$ to $(0,0)$; therefore, given that critical points are either saddles or unstable nodes, it is necessary that the origin be a saddle point. By Lemma 1, $g(\bar{s}) < 1$ is a necessary condition for a C-wave.

Assume now that $g(s) > g(\bar{s})$ on $[0, \bar{s})$. Then $g'(\bar{s}) \leq 0$, so the critical point $(\bar{u}, \bar{s})$ is an unstable node. By Lemma 66 we also have that a unique trajectory enters the origin from the first quadrant as $z \to \infty$. By Lemma 67 the boundary of the rectangle Q consists of egress points. Consequently, there is a unique orbit connecting $(\bar{u}, \bar{s})$ to $(0,0)$.

Suppose instead that there is some s in the interval $[0, \bar{s})$ such that $g(s) = g(\bar{s})$. Let s^* be the smallest such value. If $s^* = 0$, then both $(0,0)$ and $(\bar{u}, \bar{s})$ are unstable nodes, and no trajectory can connect them. Hence, assume $s^* > 0$. By Lemma 66 the boundary of the rectangle Q consists entirely of egress points. It is impossible for a trajectory to connect $(\bar{u}, \bar{s})$ to $(0,0)$ because such a trajectory would have to enter the rectangle Q in order to reach the stable manifold in the first quadrant. So we have proved the following theorem.

Theorem 68 *The system (4.21)–(4.22) has a unique C-wave solution connecting* $(\bar{u}, \bar{s})$ *to* $(0,0)$, *satisfying (4.20), if, and only if,* $g(s) > g(\bar{s})$ *for* $0 \leq s < \bar{s}$.

A remediation wave satisfies the boundary conditions

$$u(-\infty) = 0,\ s(-\infty) = 0,\ u(+\infty) = \overline{u},\ s(+\infty) = \overline{s}$$

for some $(\overline{u}, \overline{s})$ satisfying (4.20). The argument for the existence of such waves is roughly analogous to that of C-waves. We have the following theorem.

Theorem 69 *The system (4.21)–(4.22) has a unique R-wave solution connecting $(0,0)$ to $(\overline{u}, \overline{s})$ satisfying (4.20), if, and only if, $g'(\overline{s}) > 0$ with $g(s) < g(\overline{s})$ for $0 < s < \overline{s}$.*

The sufficiency of the stated conditions is demonstrated in the same way as in Theorem 68. The necessity requires a small modification. If the conditions are not met, then either $g'(\overline{s}) \leq 0$ or there exists some s in $(0, \overline{s})$ such that $g(s) = g(\overline{s})$. In the former case, $(\overline{u}, \overline{s})$ is an unstable node and there cannot be a trajectory connecting that point. In the latter case, let s^* be the largest $s < \overline{s}$ such that $g(s) = g(\overline{s})$, and let $R = \{u^* < u < \overline{u},\ s^* < s < \overline{s}\}$. By Lemma 67 the boundary of R consists of egress points and thus no trajectory can enter R from $(0,0)$.

For illustration purposes, consider the accretion function

$$F(s) = 1 + 6s - 3s^2.$$

This choice leads to all the cases discussed above, depending upon the parameter values. There are two degrees of freedom in the problem, the choice of the dimensionless paramenter b and the choice of the equilibrium state $\overline{u}$. Figure 4.1 shows the parameter space with b on the horizontal axis and u on the vertical axis. The set of possible choices of u is bounded above both by 1 and also by $b/F(1)$. Hence, there is a triangular region in the upper left portion of the plot where there are no equilibrium solutions. Theorem 68 gives conditions when a C-wave exists. Here, g'' is negative, so the condition given in Theorem 68 corresponds to $g(\overline{s}) < 1$. The condition is met for $\overline{s} > (6 - b)/3$. Because $\overline{u}$ increases with $\overline{s}$, a lower bound for s corresponds to a lower bound for u. We have the result that a C-wave propagates if, and only if,

$$\overline{u} > \frac{6b - b^2}{3 + 6b - b^2}.$$

Note that this condition is in addition to the requirements for the existence of an equilibrium state with $\overline{u} < 1$. Hence, a C-wave occurs in the region shown in figure 4.1. The conditions for an R-wave, as stated in Theorem 69, force g to be increasing on $(0, \overline{s})$, because g is concave downward. This occurs whenever $\overline{s}$ is less than $(6 - b)/6$, where g achieves it maximum. Again, a bound on $\overline{s}$ corresponds to a bound on $\overline{u}$, and we have that an R-wave occurs if, and only if,

$$\overline{u} < \frac{12b - 2b^2}{48 - b^2}.$$

This region is shown in figure 4.1. This example illustrates that some equilibrium

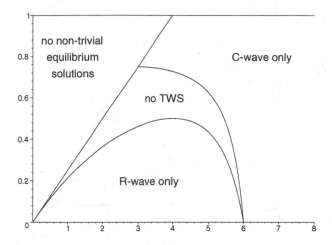

Figure 4.1: Regions in the bu parameter space corresponding to wave front solutions.

states cannot be connected to the origin by a wave front solution regardless of which of the equilibrium states is placed at infinity, while other equilibrium states do correspond to a wave front solution, provided the choice of which equilibrium state is placed at infinity is correctly made. In general, conditions giving C-waves and R-waves are mutually exclusive, so it is not possible for an equilibrium state to give both. Figures 4.2, 4.3, and 4.4 show phase portraits with $b = 4$, $a = 0.2$. Figure 4.2 uses $\bar{u} = 0.8$ and illustrates a phase portrait with a C-wave. Figure 4.3 uses $\bar{u} = 0.4$ and illustrates an R-wave. Finally, figure 4.4 shows the case $\bar{u} = 0.6$ when neither a C-wave nor an R-wave exists.

Exercise 70 *Consider the system*

$$(\epsilon\omega u + s)_t = au_{xx} - u_x,$$
$$\omega = 1 - \beta s,$$
$$s_t = F(s)u,$$

where $\epsilon, a, \beta > 0$, and F is a positive, smooth, decreasing function on $[0, s^*)$ that satisfies the conditions

$$F(0) = 1, \ F(s^*) = 0.$$

Show that there exists a traveling wave solution of speed

$$c = \frac{1}{s^* + \epsilon(1 - \beta s^*)}$$

that satisfies the boundary conditions

$$u(-\infty) = 1, \ s(-\infty) = s^*, \ u(+\infty) = s(+\infty) = 0.$$

[see Cohn, Ledder, and Logan (2000)].

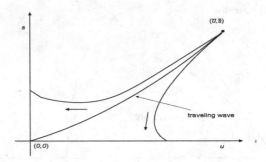

Figure 4.2: Phase portrait of a C-wave with $a = 0.2, b = 4$, and $\overline{u} = 0.8$.

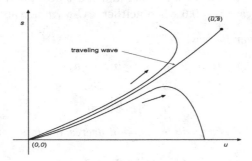

Figure 4.3: Phase portrait of an R-wave with $a = 0.2, b = 4$, and $\overline{u} = 0.4$.

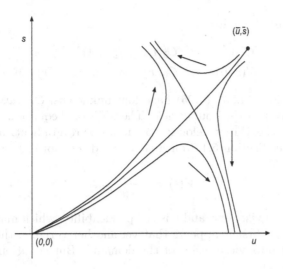

Figure 4.4: Phase portrait showing no traveling waves. Here $a = 0.2, b = 4$, and $\bar{u} = 0.6$.

4.2.4   Bounded Domains

If we limit the domain to an bounded interval, then we can better understand how boundary conditions are coupled to the flow. Again we take the one-dimensional model equations of Section 4.2, with the modification that the rate law is some general function of the concentrations and the Darcy velocity. Reiterating, the physical model is a horizontal tube in $0 \leq x \leq L$ filled with a porous material that is fully saturated. The flow is driven by maintaining a pressure gradient across the length of the tube, say, by filling a water column coupled with the tube at the left boundary. As before, C is the concentration of the solids in solution, σ and is the concentration of the retained solids. We have

$$(\omega C + \sigma)_t = \alpha V(t) C_{xx} - V(t) C_x, \qquad (4.23)$$
$$\omega = \omega_0 - \sigma, \qquad (4.24)$$
$$\sigma_t = F(C, \sigma, V(t)), \qquad (4.25)$$

where we have assumed $\omega D = \alpha V(t)$ and have neglected molecular diffusion; the function F defining the attachment-detachment mechanism is given. We take the boundary conditions to be of the form

$$p(0, t) = p_b(t), \quad p(L, t) = 0,$$

where $p_b(t)$ is a given boundary pressure. We may always assume that the right boundary is at atmospheric pressure. The volumetric concentrations satisfy the boundary and initial conditions

$$C(0,t) = C_b(t), \quad C_x(L,t) = 0,$$
$$C(x,0) = C_0(x), \quad \sigma(x,0) = \sigma_0(x).$$

The three equations (4.23)–(4.25) have four unknowns, C, σ, ω, and $V(t)$, and the pressure is given at the boundaries. The additional equation is a constitutive law that relates the Darcy velocity to the pressure gradient, and it is called **Darcy's law** (see Section 5.1.2 for a thorough discussion). The equation is

$$V(t) = -\frac{k(\omega)}{\mu}p_x, \tag{4.26}$$

where μ is the fluid viscosity, and k is the permeability, which may be a function of the porosity. At first it appears that yet another variable, the pressure, has been introduced into the interior of the domain. But equation (4.26) can be integrated to obtain

$$V(t) = \frac{p_b(t)}{\int_0^L R(\omega(x,t))dx}, \tag{4.27}$$

where

$$R(\omega) = \frac{\mu}{k(\omega)}.$$

Here, R can be interpreted as a resistance of the medium. Equation (4.26) gives the velocity in terms of the pressure on the boundary. Now, when (4.27) is substituted into (4.23)–(4.25), we obtain a nonlocal, nonlinear system for C and ω (clearly, σ can be eliminated from the equations). Numerical procedures are required to produce solutions.

The question of the dependence of permeability on porosity has led to numerous experiments to discover realistic relationships. One of the best known empirical results is the **Koseny–Carman formula**

$$k = k_0 \frac{\omega^3}{(1-\omega)^2},$$

where k_0 is a constant related to the surface area of the porous fabric that is exposed to the fluid. There are other possibilities as well [see de Marsily (1986), p. 62]. Some studies take k to be a function of the concentrations C and σ, yet take the porosity to be approximately constant.

The form of the kinetics law (4.25) is the part of the model with the most uncertainty. The rate function F may include terms that model the removal of attached particles or the accretion of suspended particles. In general, there are no theoretical-based results that fix the form of F, and therefore most models are empirical. If the Darcy velocity appears as a multiplicative factor in F, then $V(t)$ can be removed from equations (4.23)–(4.24) by a transformation, as in Section 4.2.1.

Another simplification is possible if the rate function F does not depend on the solute concentration C, i.e., $F = F(\sigma, V(t))$, and the initial concentration of retained solids is $\sigma_0(x) = \sigma_0 =$ constant. An example is the removal process defined by the rate law

$$\sigma_t = bV(t)(f(V(t)) - \sigma)_+,$$

where $(\cdot)_+ = \max\{\cdot, 0\}$ is the positive part of an expression. Then σ, and thus ω, depend only on t, and not x. Therefore (4.27) becomes

$$V(t) = \frac{p_b(t)}{LR(\omega(t))},$$

and we have a nonlinear, local model of the form

$$
\begin{aligned}
\omega' &= G(t, \omega), \\
(\omega(C - 1))_t &= (LR(\omega))^{-1} p_b(t)(\alpha C_{xx} - C_x),
\end{aligned}
$$

where G is a known function of ω and t.

Another strategy is to completely discard the dispersion term and investigate the advection model

$$
\begin{aligned}
(\omega C + \sigma)_t &= -V(t)C_x, \\
\omega &= \omega_0 - \sigma, \\
\sigma_t &= F(C, \sigma, V(t)), \\
V(t) &= \frac{p_b(t)}{\int_0^L R(\omega(x, t))dx},
\end{aligned}
$$

where $0 < x < L$ and $t > 0$. Analysis of this nonlocal model is in preparation [Wolesensky and Logan (2001)].

A simple example shows some of the difficulties encountered in analyzing nonlocal models. Consider the model system

$$
\begin{aligned}
u_t &= -v(t)u_x, \quad 0 < x < 1, \ t > 0, & (4.28) \\
u(x, 0) &= f(x), \quad 0 < x < 1, & (4.29) \\
u(0, t) &= 0, \quad t > 0, & (4.30) \\
v(t) &= \int_0^1 u(x, t)dx, & (4.31)
\end{aligned}
$$

where f is a given nonnegative, smooth function with $f(0) = 0$. Note that $v(t)$, the signal speed, is not known *a priori*, but it depends upon total mass in the system at time t. We can attempt an analytic solution using the method of characteristics (see Section 2.6). Defining characteristic curves as solutions of the differential equation

$$\frac{dx}{dt} = v(t),$$

we have, on those curves,

$$\frac{du}{dt} = u_t + u_x\frac{dx}{dt} = u_t + u_x v(t) = 0.$$

Therefore u is constant on the characteristics. By simple integration,

$$x = \varphi(t) = \int_0^t v(s)ds$$

is the equation of the leading characteristic emanating from the origin.   See figure 4.5.   In the region $\varphi(t) < x < 1$ (i.e., ahead of the leading wave), the characteristic emanating from $(\xi, 0)$ to (x, t) can be written

$$x = \varphi(t) + \xi,$$

and in the region $0 < x < \varphi(t)$ (i.e., behind the leading wave) the characteristic emanating from $(0, \tau)$ to (x, t) can be written

$$x = \int_\tau^t v(s)ds = \varphi(t) - \varphi(\tau).$$

Behind the wave, in $0 < x < \varphi(t)$, the solution is $u(x, t) = u(0, \tau) = 0$, where $\tau = \tau(x, t)$ is the solution to $x = \varphi(t) - \varphi(\tau)$. Because u is constant on the characteristic connecting $(\xi, 0)$ to (x, t), we must have $u(x, t) = u(\xi, 0) = f(\xi)$, $\varphi(t) < x < 1$. Thus,

$$u(x, t) = f(x - \varphi(t)), \quad \varphi(t) < x < 1.$$

But the characteristic speed $v(t)$, and therefore $\varphi(t)$, is not yet determined. To find an equation for $v(t)$ we write (4.31) as

$$v(t) = \int_{\varphi(t)}^1 u(x, t)dx,$$

and we differentiate, using Leibniz rule, to get

$$\begin{aligned}
v'(t) &= \int_{\varphi(t)}^1 u_t(x, t)dx - u(\varphi(t), t)\varphi'(t) \\
&= -v(t) \int_{\varphi(t)}^1 u_x(x, t)dx - u(\varphi(t), t)\varphi'(t) \\
&= -v(t)u(1, t) \\
&= -v(t)f(1 - \varphi(t)).
\end{aligned}$$

This is a differential–integral equation for $v(t)$, and the initial condition is $v(0) = v_0 = \int_0^1 f(r)dr$. But, because $v = \varphi'$, we can write

$$\varphi'' = -\varphi' f(1 - \varphi), \tag{4.32}$$

which is a second-order differential equation for φ.

Because this equation does not contain the independent variable t explicitly, we can proceed by letting $w = \varphi'$. Whence

$$\varphi'' = w' = \frac{dw}{d\varphi}\frac{d\varphi}{dt} = w\frac{dw}{d\varphi}.$$

Then (4.32) becomes

$$\frac{dw}{d\varphi} = -f(1 - \varphi).$$

Therefore, integration yields

$$w = w(0) - \int_0^\varphi f(1 - r)dr,$$

where $w(0) = \varphi'(0) = v(0) = v_0$. This equation defines w as a function of φ. Now, φ can be determined as a function of t by quadrature from $\varphi' = w$, with the initial condition $\varphi(0) = 0$.

For example, if $f(x) = x$, then $v_0 = \frac{1}{2}$, and

$$w = \frac{1}{2} - \int_0^\varphi (1 - r)dr = \frac{1}{2} + \frac{1}{2}\varphi^2 - \varphi.$$

Consequently,

$$\int \frac{d\varphi}{\frac{1}{2} + \frac{1}{2}\varphi^2 - \varphi} = t + C.$$

The integration on the left-hand-side is easily performed. When we solve for φ we obtain

$$\varphi(t) = \frac{t}{t + 2}.$$

Therefore the characteristic speed is

$$v(t) = \frac{2}{(2 + t)^2}.$$

The solution ahead of the leading signal, which, by the way, is asymptotic to $x = 1$, is

$$u(x, t) = x - \frac{t}{t + 2}, \quad \frac{t}{t + 2} < x < 1.$$

4.3 Reference Notes

The basis for the discussion in Section 4.2 is the paper by Ledder and Logan (2000). The lecture notes by Espedal, *et al.* (2000) and the monograph edited by Fasano (2000) discuss applications to the formation of industrial composites and to the seepage of fluids into hydrophilic granules; they also contain articles on numerical methods and homogenization, as well as a large number references to complex flow processes.

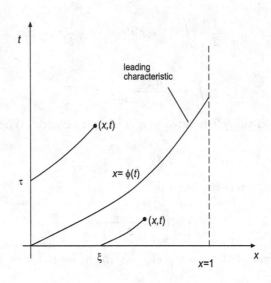

Figure 4.5: Characteristic diagram showing the leading characteristic $x = \varphi(t)$.

Chapter 5

Subsurface Flow Dynamics

Transport models in the preceding chapters focused on contaminant, or solute, transport. We derived basic conservation laws that governed how the solutes were advected and dispersed throughout the medium, and we imposed relations that governed the kinetics of adsorption of those solutes. The subject of contaminant transport is just one aspect of porous media flow, and the development of the theory of these processes is fairly recent in the overall history of the study of ground water processes, beginning roughly in the last third of the twentieth century. Long before contaminants were of interest, researchers were interested in how the groundwater itself behaved as it seeped through porous structures. Questions involving water levels in wells, springs, and reservoirs, seepage through earthen dams, and the effects of pumping water out of wells occupied the early investigators, even back in the mid-nineteenth century.

5.1   Darcy's Law

Heretofore we superimposed a constant velocity flow upon the porous domain and we paid little attention to the causes, i.e., the driving forces behind the flow. The equations coming out of the analysis were the solute transport equation, or mass balance, and the chemical rate laws. Thus, we did some of the kinematics and chemistry, but we ignored the dynamics of the underlying flow. Now we wish to broaden our perspective and consider velocity changes, porosity changes, and other interconnections between the flow variables, disregarding the fact that solutes are present. Therefore, we now focus attention on the water that flows through the medium.

Understanding the driving forces of subsurface water (groundwater) is of immense practical importance in the management of water resources. The causes and effects of water table fluctuations, changes in aquifer and well levels, and the effects of irrigation and droughts all are part of a large body of applied science termed groundwater hydrology or **hydrogeology**.

Hydrogeology is one of the three basic elements of the water cycle, the

other two being meteorology (clouds, rain, atmospheric water) and surface water (rivers, streams, dams, floods, runoff).   Subsurface water analysis is difficult because of its variability.   We cannot see underground, and extensive
measurements and observations are difficult.   The medium itself is extremely
heterogeneous, and analysis often involves several scales. **Aquifers**, which are
subsurface domains that are able to store and transmit water, may be on the
scale of 10^4 meters or more, and large-scale heterogeneities within an aquifer
may range from 10^{-1} to 10^2 meters. The pores themselves may be on the order
of 10^{-4} meters, and if adsorption and chemical processes are of interest, then
analysis may require looking on the order of 10^{-7} meters, which is the order
of the adhesive layer on the soil fabric.   Then add to these length scales the
possible time scales.   Water may flow meters per day, yet chemical reactions
may occur rapidly and mineral deposition may take centuries.   The mix of all
these length and time scales makes problems difficult to formulate and solve.

We usually model aquifers by considering two regions.   Directly below the
soil surface is the **unsaturated zone,** which contains both air and water (or
moisture); not all of the pore space is filled with water and gravity and capillary forces are important.   As we move down in elevation we encounter the
saturated zone, where all the pore space is filled with water.   Separating
these two zones is a narrow region called the **water table** where the pressure
is atmospheric. The water table is modeled as a surface.

Most of the analysis in our subsequent discussion occurs below the water
table, in the saturated zone.   There, the basic idea is that the subsurface fabric
structure (soil, sand, rock) is a porous medium through which water flows,
caused by excessive hydrostatic pressures, which are in turn caused by irrigation,
precipitation, pumping from wells, and so on.   The first step in the analysis
is to relate the flow rates, determined by the velocity field, to the pressure
gradients.   The result is Darcy's law, the most fundamental phenomenological
law in hydrogeology, which posits the mechanism for the flow in the saturated
zone.

5.1.1   Kinematical Flow Relations

Consider an arbitrary volume Ω in a three-dimensional porous domain where
a fluid of density $\rho = \rho(x,t)$ is flowing with Darcy velocity $V = V(x,t)$, and
suppose that the porosity is evolving in both space and time, i.e., $\omega = \omega(x,t)$,
where $x \in \mathbb{R}^3$. We assume the flow is saturated, i.e., the pores are completely
filled with water.   Then the total amount of fluid in Ω at any instant of time
t is $\int_\Omega \rho\omega dx$, and the rate that fluid flows through an oriented surface element
ndA is $-\rho V \cdot ndA$. Therefore, if there are no fluid sources in the region, we can
balance the mass of fluid in Ω to conclude that

$$\frac{d}{dt} \int_\Omega \rho\omega dx = - \int_{\partial\Omega} \rho V \cdot ndA.$$

With an application of the divergence theorem over the abitrary domain Ω, we obtain

$$(\rho\omega)_t + \nabla \cdot (\rho V) = 0,\qquad (5.1)$$

which is the local **fluid mass balance law**.

To analyze this equation more carefully, let us expand the derivatives to obtain

$$\rho\omega_t + \omega\rho_t + \rho\nabla \cdot V + V \cdot \nabla\rho = 0.$$

If we introduce the logarithmic density $\delta = \ln(\rho/\rho_0)$, where ρ_0 is a constant reference density, we can write the preceding equation in the form

$$\nabla \cdot V = -\omega_t - (\omega\frac{\partial}{\partial t} + V \cdot \nabla)\delta.\qquad (5.2)$$

With the usual calculus interpretation of the divergence, the right-hand side has two terms that can be regarded as source terms for the fluid velocity. If the density is constant (which, for most of the time in the sequel, will be our assumption), then the equation leads to the **continuity equation**

$$\omega_t + \nabla \cdot V = 0,$$

which relates porosity changes to velocity changes. The one-dimensional version of the continuity equation is

$$\omega_t + V_x = 0.$$

Generally, the first term on the right-hand side of (5.2) denotes porosity changes. If the porosity is decreasing, then fluid is expelled from the pores causing an increase in the divergence of the flow. Unless there are sudden geologic structural changes, one might expect porosity changes to only occur over long time scales, and thus the rate of porosity changes will be small. The second term on the right of (5.2) involves the changes, both temporal and spatial, of density. If v_0 is a velocity scale for the flow and L is a length scale, then the divergence term on the left is order v_0/L, while the term $V \cdot \nabla\delta$ is order v_0/L times the order of the density changes, which are often small. If both terms on the right-hand side of (5.2) can be ignored, then we have

$$\nabla \cdot V = 0,$$

which is the **incompressibility condition**. For most flows in porous media, it is safe to assume this condition. This was the basic assumption in preceding chapters. This condition occurs, of course, when the density of the water and the porosity are constants.

If there are chemical reactions between the fluid and mineral rock and dehydration occurs, then a source term must be added to the right-hand side of (5.1) which represents the rate of generation of water per unit volume. In this case the fluid mass balance equation becomes

$$(\rho\omega)_t + \nabla \cdot (\rho V) = \rho q,\qquad (5.3)$$

where q is the rate of generation.

5.1.2   Darcy's Law in One Dimension

Classical fluids are governed by mass balance and momentum balance, the latter expressed by the Navier–Stokes equations

$$\rho(u_t + (u \cdot \nabla)u) = \rho f + \mu \Delta u - \nabla p,$$

where u is the actual vector fluid velocity in the pores, ρ is the density, μ is the viscosity, p is the pressure, and f is the body force (e.g., gravity). If there are temperature changes, then energy balances, as well as equations of state, must be included. In porous media, however, the stress and velocity fields through the many, narrow porous pathways are complicated, and the Navier–Stokes equations are not readily adaptable. Therefore, in porous media the dynamical equations are usually replaced by a constitutive equation based on experimental results relating the discharge of the liquid through the medium. This constitutive relation is Darcy's law, and it is based on the empirical results of Darcy in the mid 1800s on the relation between pressure changes and flow velocity through columns filled with porous material. Darcy's law confirms our intuition that the flow velocity is driven by pressure and elevation changes across the medium.

In the simplest case, Darcy's law in a one-dimensional porous medium states that the discharge Q at x, that is, the amount (volume) of water traversing the cross-section of a tube at x, per unit time, is directly proportional to the pressure gradient at that point and to the cross-sectional area A. In symbols,

$$Q(x,t) = -AK\frac{1}{\rho g}p_x(x,t),$$

where $p = p(x,t)$ is the pressure and $\rho = \rho(x,t)$ and g are the density of the fluid and the gravatational constant, respectively. The proportionality factor K is called the **hydraulic conductivity** and has dimensions of velocity. The discharge per unit area, that is Q/A, is the **Darcy velocity**, or **filtration velocity**. If ρ is constant, we introduce the quantity $h(x,t) = p(x,t)/\rho g$, which is called the **head** and has dimenions of length; then **Darcy's law** in one dimension can be written

$$V(x,t) = -Kh_x(x,t), \quad h = \frac{p}{\rho g}.$$

In words, the fluid velocity is driven by pressure gradients. The hydraulic conductivity is a measure of the medium's ability to conduct the water; it may not be constant, but depend upon position and other quantities in the problem. (It is often easier to understand the definition of the head in three dimensions, which will be the subject of the next section.)

Observe that Darcy's law is similar to other common laws in physics. It is analogous to Fick's law, which states that the flux of chemical species is proportional to its concentration gradient ($\phi = -DC_x$, where D is the diffusion constant), or Fourier's heat flow law, which states that the heat energy flux is proportional to the temperature gradient ($\phi = -KT_x$, where K is the thermal conductivity); in electrodynamics, one often assumes Ohm's law, which states that the current density is proportional to the gradient of the potential ($J = -\sigma U_x$, where σ is the electrical conductivity and U is the potential). These are all constitutive relations that are based on empirics. So it is with Darcy's law; it is a phenomenological result that is the basic constitutive relation in hydrogeology.

As we noted, the hydraulic conductivity is not generally a constant. It can depend on position for a nonhomogeneous medium, and it can (and will) depend on porosity. Because the conductivity varies inversely with the viscosity of the fluid, we often replace K by the ratio

$$k = \frac{\mu K}{\rho g},$$

where μ is the dynamic viscosity (mass length^{-1} time^{-1}) and k is the **permeability**, given in units of length^2. Then Darcy's law in one dimension is

$$V = -\frac{k}{\mu}p_x. \tag{5.4}$$

This is the general form of Darcy's law, holding in the case of either constant or nonconstant density. We also assume its validity for both static and time-dependent flows.

5.1.3   Darcy's Law in Higher Dimensions

Darcy's law in three-dimensional porous domains includes two contributions to the vector Darcy velocity. One is the pressure gradient, and the other, not present in one dimension, is the elevation. We first fix a level $z = 0$ reference plane; for example, this level could be at the bollom of an aquifer. Then, at the point (x, y, z) we define the **head** $h = h(x, y, z, t)$, a scalar function with dimensions of length, by the formula

$$h(x, y, z, t) = z + \frac{1}{\rho g}p(x, y, z, t), \tag{5.5}$$

where p is the pressure. The two terms on the right-hand side are called the **elevation head** and the **pressure head**, respectively. To understand the head, or hydraulic head, h, which is used extensively in hydrology, let us hold time fixed and think quasistatically. Take the the porous medium to be bounded below by an impermeable layer at a reference plane $z = 0$, and imagine that an open tube is inserted from the top and has its lower end at a point $P : (x, y, z)$ in the water-saturated medium. The head at P is the height above the reference plane $z = 0$ to which a column of water will rise in the tube. (Such tubes are called piezometers, standpipes, or manometers.) See figure 5.1. In an unconfined aquifer, where the subsurface water is in free communication with the atmosphere, the water column in the tube will rise to the free surface of the

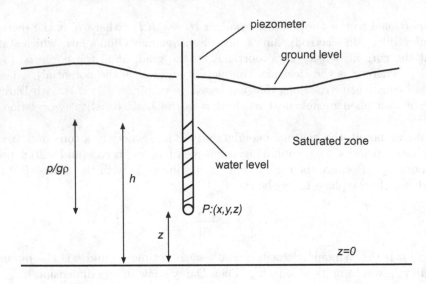

Figure 5.1: Open tube with the lower end at the point $P : (x, y, z)$. The head at P is the height above $z = 0$ to which a column of water will rise in the tube.

water (the water table); in a confined aquifer, which is bounded above by an impermeable or confining layer, the water below is under pressure and the water column could rise to a level exceeding the elevation of the top of the aquifer or even above the ground. So, the head is the distance z above the reference level plus the distance $p(x, y, z)/\rho g$, this latter being the length of the water column in the tube from z to the surface (recall, by hydrostatics, the pressure is ρg times the depth). In the case of an unconfined aquifer, if the water table is at $z = H(x, y)$, then the pressure is atmospheric (taken to be $p = 0$), and the head at the free surface is just the distance from the bottom reference level; the head at bottom is $h = p(x, y, 0)/\rho g$, where $p(x, y, 0)$ is the pressure at the bottom.

Darcy's law in three dimensions states that the filtration velocity is "down the pressure gradient," with an elevation contribution; in symbols

$$V = -\frac{k}{\mu}\nabla(p + \rho g z), \tag{5.6}$$

where k is the permeability and μ is the viscosity. We assume Darcy's law holds for time-dependent flows as well. When the density ρ is constant, it is common to write Darcy's law in terms of the head as

$$V = -K\nabla h, \quad h = z + \frac{1}{\rho g}p, \tag{5.7}$$

where K is the hydraulic conductivity. In general, the hydraulic conductivity

K (or equivalently, the permeability k) can be a tensor (matrix) of the form

$$K = \begin{pmatrix} K_{11} & K_{12} & K_{13} \\ K_{21} & K_{22} & K_{23} \\ K_{31} & K_{32} & K_{33} \end{pmatrix}.$$

In a heterogeneous medium, the components can be spatially dependent, or they can be dependent on other quantities in the problem. In the general case the Darcy velocity component in a given direction will depend on all three components of the head gradient. But in many instances K is a diagonal matrix so that the components of V are given by $V_i = K_{ii}\partial h/\partial x_i$, where $(x, y, z) = (x_1, x_2, x_3)$. In this case the conductivity of the medium may be different in the three coordinate directions, as in a stratified medium. In the most general case the conductivity could have any preferred directions and one could transform to coordinates in the principal directions to analyze these cases; these calculations are beyond our scope and we refer the reader to the books listed in the references [e.g., see Bear (1988)].

There have been many efforts to justify or prove Darcy's law, based on the Navier–Stokes equations. One simple, intuitive explanation is as follows. As water moves through the pores of the medium, there is a resistance caused by fluid viscosity and the small dimenions of the pores. The Navier–Stokes equations have the form

$$\rho\frac{Du}{Dt} = -\nabla p - \rho g\mathbf{k} + \text{viscous force},$$

where ρ is the fluid density, u is the vector velocity, $Du/Dt = u_t + (u \cdot \nabla)u$ is the material derivative, and the body force is $-g\mathbf{k}$ (here, $\mathbf{k}$ is the unit vector in the z-direction); the viscous force is $\mu\Delta u$. If the inertial term on the left-hand side is small, then we have, approximately,

$$\text{viscous force} = \nabla p + \rho g\mathbf{k}.$$

We might also intuit that the viscous force, which is similar to a resistive force, should be proportional to the filtration velocity V; thus,

$$V = -\alpha(\nabla p + \rho g\mathbf{k}),$$

which is of the form of Darcy's law. One reason for introducing the head becomes clear when just looking at the form of the Navier–Stokes equations.

A better justification can be given by imagining that the medium is a type of capillary system; that is, it has cross-sectional area A and the pores are n uniform, cylindrical tubes, or ducts, of radius R and length L (see figure 5.2).

The flux through a single duct can be found by solving the Navier–Stokes equations, and then we can average to get the total flux. For readers familiar with fluid mechanics, this is Poiseuille flow. We impose a cylindrical coordinate system in the cylindrical duct with the z axis in the axial direction and r

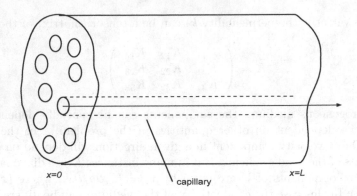

Figure 5.2: A one-dimensional porous medium modeled as a system of parallel, cylindrical ducts.

measured radially from the axis. In a single duct, in the absence of inertial or gravitational forces, the Navier–Stokes equations are

$$\nabla \cdot u = 0, \quad \nabla p - \mu \Delta u = 0,$$

where μ is the viscosity. Along the lateral sides of the cylinder $(r = R)$ we assume that the velocity is zero and at the ends of the duct we assume the pressure is given by $p = p_1$ at $z = 0$ and $p = p_2 < p_1$ at $z = L$. We further assume that the velocity has only an axial component, or $u = (u^{(r)}, u^{(\theta)}, u^{(z)}) = (0, 0, u^{(z)})$. The incompressibility condition becomes simply $u^{(z)} = u^{(z)}(r)$, or the velocity component in the z direction depends only on the radial coordinate r. In the r and θ directions the pressure equation forces p to be independent of r and θ. In the z direction the equation becomes

$$p'(z) = \mu \left(\frac{d^2 u^{(z)}}{dr^2} + \frac{1}{r} \frac{du^{(z)}}{dr} \right).$$

The left-hand side of this equation depends only on z, and the right-hand side depends only on r. This fact forces both sides to be constant and so we have $p(z) = az + b$, where a and b are constants. Applying the two boundary conditions on pressure, we obtain

$$p(z) = p_1 + \frac{p_2 - p_1}{L} z.$$

The velocity equation then becomes

$$\frac{d}{dr} \left(r \frac{du^{(z)}}{dr} \right) = \frac{p_2 - p_1}{L} \frac{r}{\mu}.$$

Integrating this equation twice, while using the boundary condition $u^{(z)}(R) = 0$ (this is the adherence boundary condition of viscous flow), as well as the boundedness of the velocity, we get a radial, parabolic velocity profile

$$u^{(z)}(r) = \frac{p_2 - p_1}{4L\mu}(r^2 - R^2).$$

This type of flow is called **Poiseuille flow**, and a thorough analysis can be found in any fluid mechanics book. Knowing the velocity profile, we can calculate the flux through the single duct. We have

$$\text{flux} = q = \int_0^R \int_0^{2\pi} \frac{p_2 - p_1}{4L\mu}(r^2 - R^2)rdrd\theta = \frac{\pi(p_1 - p_2)R^4}{8L\mu}.$$

Because there are n ducts, the total flow through the medium is given by $Q = nq$. The porosity is $\omega = n\pi R^2/A$. Therefore, the total flow is

$$Q = -A\left(\frac{\omega R^2}{8}\right)\frac{1}{\mu}\frac{p_2 - p_1}{L}.$$

This compares well with Darcy's law, which states that the volumetric flow rate, or discharge, is proportional to the area and to the pressure gradient. As anticipated, the proportionality constant depends on the porosity. One can also introduce gravity into this calculation; we leave this as an exercise. There are also justifications of Darcy's law based on homogenization theory [e.g., see Fowler (1997)].

Exercise 71 *Repeat the last argument and include gravity in the calculation.*

We emphasize again that Darcy's law is a macroscopic model of the dynamics of the flow and in no way represents the flow within the individual pores. It accounts completely for viscous effects and thus further considerations of the flow need not include viscosity or friction.

Exercise 72 *In gas flow through porous rock the porosity is constant and Darcy's law holds in the form $V = -(k/\mu)\nabla p$. If the equation of state is the γ-law gas equation $p = p_0\rho^\gamma$, where p_0 is constant and $\gamma > 1$, show that the density satisfies the porous media equation*

$$\rho_t = \alpha\Delta(\rho^m)$$

for some constants $\alpha > 0$ and $m > 2$. If $x \in R^n$ and $\alpha = 1$, derive the Barenblatt solution

$$\rho(x,t) = t^{-a}\left(c - \frac{a(m-1)}{2nm}\frac{|x|^2}{t^{2a/n}}\right)^{1/(m-1)}_+$$

by assuming a similarity solution of the form $\rho(x,t) = t^{-a}f(|x|/t^b)$. Here $a = (m-1+2/n)^{-1}$, $(\cdot)_+ = \max\{0,\cdot\}$, and $|x|$ is the distance from the origin. The constant c characterizes the total mass

$$\int_{R^3} \rho(x,t)dx = constant.$$

In one dimension with $m = 2$ and $c = 1/12$ find the solution and graph density profiles for several different times.

5.2   Models of Groundwater Flow

5.2.1   The Governing Equations

The basic model equations of hydrogeology are mass balance and Darcy's law, the latter providing the dynamical relation. In three dimensions we have [see (5.3) and (5.6)]

$$(\rho\omega)_t + \nabla \cdot (\rho V) = \rho q, \tag{5.8}$$

$$V = -\frac{k}{\mu}\nabla(p + \rho gz). \tag{5.9}$$

Thus, there are four equations in six unknowns: p, ρ, ω, and the three components of V. Therefore we need additional assumptions, in the way of one or more constitutive relations, to complete the system.

One possible assumption is to take constant density. Then the system (5.8)–(5.9) becomes

$$\omega_t + \nabla \cdot V = q,$$

$$V = -\frac{k}{\mu}\nabla(p + \rho gz).$$

Introducing the conductivity and eliminating V yields

$$\omega_t = \nabla \cdot (K\nabla h) + q. \tag{5.10}$$

This single equation relates porosity changes to head changes, or pressure changes, in the medium, and there remains two unknowns.

One direction to go at this point is toward **consolidation theory**. Here, the fabric compacts under its own weight and expels the water. Completion of the consolidation problem requires a constitutive relation between the pressure in the fluid and the porosity to supplement (5.10). Generally, observations show that increasing water pressure causes an expansion of the mineral or soil fabric with a corresponding decrease in porosity; a decreasing pressure leads to a contraction of the fabric and increased porosity. A detailed discussion of consolidation can be found in deMarsily (1986).

Another direction is to assume both the density and porosity are constant. Then we have $\nabla \cdot V = q$; this condition combines with Darcy's law to give the single equation

$$-\nabla \cdot (K\nabla h) = q. \tag{5.11}$$

In the case of constant conductivity, equation (5.11) becomes

$$-\Delta h = \frac{q}{K},$$

which is **Poisson's equation** for the head. If there are no sources, then we have **Laplace's equation**

$$\Delta h = 0.$$

If $K = \text{diag}(K_1, K_2, K_3)$ is diagonal, then (5.11) becomes

$$(K_1 h_x)_x + (K_2 h_y)_y + (K_3 h_z)_z = -q.$$

This elliptic equation is fundamental in groundwater studies. Its analysis forms a large part of the subject, along with the many applications to specific systems.

Exercise 73 *Find the general solution $h = h(x)$ of the one-dimensional equation*

$$-(K(x)h'(x))' = q(x),$$

where both the hydraulic conductivity and source function depend on x. Find the solution on the interval $[0, L]$ when this equation is supplemented by boundary conditions $h(0) = h_0$ and $h(L) = h_L$.

5.2.2 Boundary Conditions and Free Surfaces

The equilibrium, or elliptic, partial differential equations governing the hydraulic head must be supplemented by boundary conditions that hold on the boundaries of the domain under consideration. There are two types of flows. One is characterized by a flow domain Ω in which one part of the boundary is a free surface. Such domains are called **unconfined domains**, in contrast to **confined domains** where all the boundaries of Ω are fixed. We refer to the flows in such domains as either unconfined flows or confined flows. The free surface (also, the piezometric surface, the phreatic surface, or the water table) is the set of those points where the pressure is atmospheric; by convention, this means zero pressure. Hence, the value of the head on the free surface is just the depth of the water at that point, provided we take the reference level $z = 0$ to be the lower boundary of the flow domain. Below the free surface the flow is saturated, which means that all the pore space is completely filled with liquid. Above the free surface is the unsaturated zone (or vadose zone) where the pores are filled mainly with air and water vapor (see Section 5.3.2). The free surface is a model of a very narrow layer that separates these two zones. In terms of the head, the free surface is characterized by the equation

$$S(x, y, z) := h(x, y, z) - z = 0.$$

The free surface is also assumed to be composed of the same fluid particles, and therefore the velocity $V = (V_1, V_2, V_3)$ is parallel to the surface, and consequently perpendicular to the normal of the surface. In symbols, $\nabla S \cdot V = 0$ on the free surface. Expanding this out we get the **free surface condition**

$$\nabla h \cdot V = V_3 \quad \text{on} \quad z = h(x, y, z).$$

Equivalently, on a free surface we must have a Neumann boundary condition of the form

$$-n \cdot K \nabla h = 0,$$

where n is the normal vector. If there is recharge (e.g., irrigation, precipation, or evaporation), then a source term r must added to the last condition to obtain

$$-n \cdot K \nabla h = r$$

on the free surface. For unconfined flows the location of the free surface must be determined as part of the solution to a problem. Boundary value problems over domains where one portion of the boundary is unknown are often called free boundary value problems. These unknown boundaries in a problem usually create more complicated analyses.

By Darcy's law the velocity field is down the gradient of the head and is therefore normal to the level surfaces $h(x, y, z) =$ constant.   Thus, the flow lines or particle paths (streamlines, in steady flow) must be orthogonal to the level surfaces of h.   By sketching a rough velocity field and the corresponding streamlines, we can get a good notion of the surfaces along which h is constant. The picture is exactly analogous to that in electrostatics where the equipotential surfaces are normal to the electric flux lines determined by the electric field vector. So the head h is a potential-like function for groundwater flow.

Boundary conditions on the head at *fixed* portions of the boundary depend upon the physical conditions there. An **impermeable boundary** is one where the fluid cannot penetrate. Thus, the velocity vector must be parallel to such a boundary and we again obtain the Neumann boundary condition

$$-n \cdot K \nabla h = 0.$$

If the boundary adjoins a reservoir of water, then the pressure distribution along the boundary can be taken to be hydrostatic. It follows that the head is constant along that boundary, giving a Dirichlet boundary condition, called a **reservoir boundary** condition:

$$h = h_0 = \text{constant}.$$

Finally, along a discontinuity surface in material properties (for example, the surface between two porous domains of different conductivities), the hydraulic head must be continuous and the flux must be continuous (the water entering into a surface element must equal the water leaving on the other side); this latter condition is clearly expressed by

$$K_- n \cdot \nabla h_- = K_+ n \cdot \nabla h_+,$$

where the subscripts on K and h denote values on the two sides of the discontinuity, and n is the normal.

Another type of boundary is a **seepage face** where water seeps outward along an outlet surface. At a fixed seepage face we have $h(x, y, z) = z$ because the surface is at atmospheric pressure, and $-n \cdot \nabla h > 0$ because the flow is outward. Seepage boundaries pose the same type of difficulties as free surfaces because the length of the seepage face, where it begins and ends, has to be determined as a part of the problem.

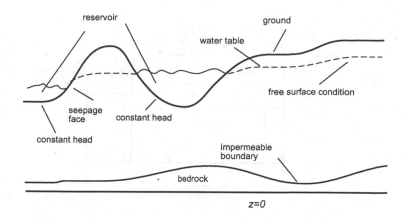

Figure 5.3: Cross section of porous medium where various boundary conditions apply to the head h.

Figure 5.3 shows a generic medium with various boundaries where different conditions hold.

We work out a simple example. Consider two reservoirs of depths h_0 and h_1 connected by a flow-through confined aquifer of length L and height b. See figure 5.4. There is no free surface in this case. The governing boundary value problem for the head $h = h(x)$ is

$$\begin{aligned}
(Kh')' &= 0, \quad 0 < x < L, \\
h(0) &= h_0, \quad h(L) = h_1.
\end{aligned}$$

If K is constant, we have the linear solution

$$h(x) = h_0 + \frac{h_1 - h_0}{L}x.$$

The flow rate per unit width is

$$\text{flow rate} = b \cdot 1 \cdot |V| = bK|h'(x)| = bK\frac{|h_1 - h_0|}{L}.$$

The product $T = bK$ is called the transmissivity. Observe that, if the aquifer is not confined, then this linear function for $h(x)$ would *not* be the solution; the upper boundary is not a no-flow boundary because it does not satisfy the free-surface condition:

$$n \cdot \nabla h = (n_1, n_2) \cdot (\frac{h_1 - h_0}{L}, 0) = n_1\frac{h_1 - h_0}{L} \neq 0.$$

Exercise 74 *A cylindrical, confined aquifer of radius R, thickness b, and constant conductivity K has a well in its center of radius r_w. The well pumps out*

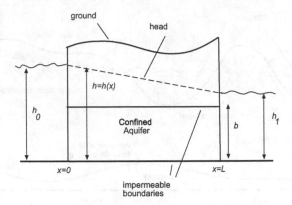

Figure 5.4: A confined, one-dimensional flow-through aquifer with head $h = h(x)$.

water at the constant rate q; the well is fully penetrating, i.e., it draws surround-ing water uniformly through its entire length. The head at $r = R$ is H, and $H > b$. Show that the head $h = h(r)$ satisfies the boundary value problem

$$\frac{1}{r}(rh'(r)) = 0, \quad r_w < r < R,$$

$$r_w h'(r_w) = \frac{q}{2\pi Kb}, \quad h(R) = H.$$

Derive the Thiem–Dupuis formula

$$h(r) = H + \frac{q}{2\pi Kb} \ln(r/R).$$

Hint: To obtain the boundary condition note that the flux through a surface element $dA = br_w d\theta$ of the well is $V dA$ where $V = -Kh'$. The total flux is q.

Exercise 75 *In the previous exercise, what happens if the aquifer is unbounded, that is, the boundary $r = R$ is moved to $r = \infty$ and the boundary condition is $h(\infty) = H$? Is there a steady solution in this case?*

5.2.3 The Dupuis Hypothesis

We have remarked that in unconfined domains the unknown position of the free surface adds a significant complication. However, if the free surface is nearly horizontal, where the head has a weak dependence on z, then an approximation can be made which greatly simplifies the analysis of these flows. These condi-tions often hold because the horizontal length of an aquifer is much larger than its depth. This approximation is called the Dupuis hypothesis, named for the French hydrologist who laid the basis for it in the 1860s.

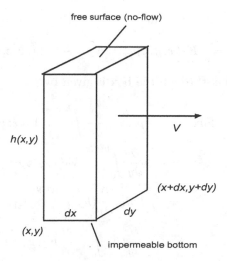

Figure 5.5: Small box with impenetrable lower boundary and upper boundary $h = h(x, y)$.

Assume the density and porosity are constant, and assume the velocity field is horizontal and parallel along each vertical line. This really means the constant head surfaces (lines in two dimensions) are nearly vertical. Away from inlets and outlets, these are often good assumptions. Thus, the velocity field is

$$V = (V_1(x, y, z), V_2(x, y, z), 0).$$

Then, by Darcy's law, h is independent of z, or

$$h = h(x, y),$$

and h represents the height of the free surface at (x, y). Note that V can vary with z; only its vertical component is zero. We now consider a small box whose base is an impermeable lower boundary and whose top is the free surface (see figure 5.5), and we balance mass flow in the box. We have

$$\text{Net flux in the } x\text{-direction} = \rho dy \int_0^{h(x,y)} V_1(x, y, z)dz$$
$$-\rho dy \int_0^{h(x+dx,y)} V_1(x, y, z)dz$$
$$= -\rho dy \left(\frac{\partial}{\partial x} \int_0^{h(x,y)} V_1(x, y, z)dz\right)dx$$

Similarly,

$$\text{Net flux in the } y\text{-direction} = -\rho dx \left(\frac{\partial}{\partial y} \int_0^{h(x,y)} V_2(x, y, z)dz\right)dy.$$

If we use Darcy's law in the form

$$V_1 = -K_1(x, y, z)h_x, \quad V_2 = -K_2(x, y, z)h_y,$$

then the net mass flux through the box is given by

$$
\begin{aligned}
\text{Net mass flux} \quad = \quad & -\rho dx dy [\frac{\partial}{\partial x} \int_0^{h(x,y)} V_1(x, y, z)dz \\
& +\frac{\partial}{\partial y} \int_0^{h(x,y)} V_2(x, y, z)dz] \\
= \quad & \rho dx dy [\frac{\partial}{\partial x}(h_x \int_0^{h(x,y)} K_1(x, y, z)dz) \\
& +\frac{\partial}{\partial y}(h_y \int_0^{h(x,y)} K_2(x, y, z)dz)].
\end{aligned}
\tag{5.12}
$$

Therefore, under steady conditions with no sources ($q = 0$) we have

$$\frac{\partial}{\partial x}\left[h_x \int_0^{h(x,y)} K_1(x, y, z)dz \right] + \frac{\partial}{\partial y}\left[h_y \int_0^{h(x,y)} K_2(x, y, z)dz \right] = 0.$$

If $K_1 = K_2 = K = $ constant, then

$$(hh_x)_x + (hh_y)_y = 0,$$

or

$$(h^2)_{xx} + (h^2)_{yy} = 0.$$

Thus we obtain Laplace's equation for the square of the head. This approximation is called **Dupuis–Forchheimer's equation**. If there is a non-zero source q, then there is a term $-q/K$ on the right side.

For an example, consider an unconfined one-dimensional porous medium connecting two reservoirs of depths h_0 and h_1. See figure 5.6. The Dupuis–Forchheimer equation for the head is $(h(x)^2)'' = 0$, whose general solution is $h(x)^2 = ax + b$. The boundary conditions $h(0) = h_0$ and $h(L) = h_1$, $h_0 > h_1$, determine the constants a and b and we easily obtain

$$h(x) = \sqrt{h_0^2 + \frac{h_1^2 - h_0^2}{L}x}.$$

Therefore, the Dupuis approximation is not a linear function. Figure 5.6 shows both the approximation and the exact solution, which includes a seepage face at the right boundary. The derivatives at $x = 0$ and $x = L$ are

$$h'(0) = \frac{h_1^2 - h_0^2}{2Lh_0}, \quad h'(L) = \frac{h_1^2 - h_0^2}{2Lh_1},$$

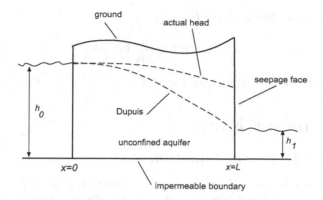

Figure 5.6: Unconfined porous medium connecting two reservoirs of depths h_0 and h_1.

and so we get inconsistencies at the two boundaries (the fluxes are not continuous). Nevertheless, we can compute the flow rate per unit width at $x = L$ using the approximation; we obtain

$$\text{flow rate} = Kh'(L)h_1 = K\frac{h_1^2 - h_0^2}{2L}.$$

This approximation is not only good, but it turns out to be exact [e.g., see Bear (1988), p. 366].

Exercise 76 *Consider the problem in the previous example and include a positive source term, i.e., a constant water accretion rate $q > 0$, say, due to rainfall from above. Show that the boundary value problem is*

$$\frac{K}{2}(h^2(x))'' = -q, \quad 0 < x < L; \quad h(0) = h_0, \quad h(L) = h_1.$$

Derive the formula for the head given by

$$h(x) = \left(h_0^2 + \frac{h_1^2 - h_0^2}{L}x + \frac{q}{K}(L - x)x\right)^{1/2}$$

and sketch the solution. How does the solution differ if we consider evaporation instead of rainfall?

Exercise 77 *A cylindrical, unconfined aquifer of radius R and conductivity K has a well in its center of radius r_w that pumps out water at the constant rate q. The head at $r = R$ is H. Show that the head $h = h(r)$ satisfies the boundary value problem*

$$\frac{1}{r}\frac{d}{dr}(r\frac{d}{dr}h^2) = 0, \quad r_w < r < R,$$

$$r_w\frac{d}{dr}h^2(r_w) = \frac{q}{\pi K}, \quad h(R) = H.$$

Derive the Dupuis–Forchheimer formula

$$h^2(r) = H^2 + \frac{q}{\pi K} \ln(r/R).$$

Comment on the error in this approximation.

5.3   Transient Flows

5.3.1   The Diffusion Equation

If transients are present, then the head is changing with time.   We analyze the case of an *unconfined* domain and make the Dupuis hypothesis; thus, $h = h(x, y, t)$. Again we consider the small box shown in figure 5.5 and balance the mass. The amount of water added per unit time is given by

$$\rho \omega_d h_t dx dy,$$

where $\omega_d < \omega$ is called the drainage porosity or **specific yield**. Here ω_d is the volume of water that drains from the saturated fabric due to gravity per total volume of the box.   We also assume that Darcy's law in the transient state is the same as in steady state.   Then the water balance equation is, from (5.12),

$$\omega_d h_t = \frac{\partial}{\partial x} \left[h_x \int_0^{h(x,y)} K_1(x,y,z)dz \right] \tag{5.13}$$
$$+ \frac{\partial}{\partial y} \left[h_y \int_0^{h(x,y)} K_2(x,y,z)dz \right].$$

If K_1 and K_2 are constant, this equation simplifies to

$$\omega_d h_t = \left(\frac{K_1}{2} h^2 \right)_{xx} + \left(\frac{K_2}{2} h^2 \right)_{yy},$$

which is the **Boussinesq equation**. It is common in hydrogeology to introduce the **transmissivities**

$$T_1(x,y) = \int_0^{h(x,y)} K_1(x,y,z)dz, \quad T_2(x,y) = \int_0^{h(x,y)} K_2(x,y,z)dz.$$

Then the time-dependent equation (5.13) can be written

$$\omega_d h_t = (T_1 h_x)_x + (T_2 h_y)_y.$$

In the special case that the transmissivities T_1 and T_2 are nearly constant, say, both equal to T, we obtain the two-dimensional diffusion equation

$$\omega_d h_t = T \Delta h. \tag{5.14}$$

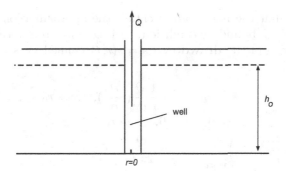

Figure 5.7: An infinite, unconfined porous medium with a fully penetrating well of radius r_0 at its center.

Equation (5.14) is the diffusion equation that is often applied in unconfined flow to calculate the evolution of the head. In this case we generally take $T = Kb$, where b is the approximate vertical thickness of the domain.

It is common to consider problems involving radial flow fields, for example, those caused by pumping water from a well. In this case we transform to cylindrical coordinates in (5.14).

We now take up an example. Consider a radially symmetric unconfined porous domain in $z, r > 0$ with a fully penetrating vertical well at $r = 0$ of small radius r_0 that is pumping out water at the rate $Q > 0$. See figure 5.7. At time $t = 0$ the head is a constant h_0, and we are interested in determining the head as a function of the radius $r > r_0$ and time $t > 0$. The initial boundary value problem is

$$h_t = a\frac{1}{r}\frac{\partial}{\partial r}(r\frac{\partial h}{\partial r}), \quad r > r_0, \quad t > 0,$$

$$h(r,0) = h_0, \quad r > r_0,$$

$$\left(r\frac{\partial h}{\partial r}(r,t)\right)_{r=r_0} = \frac{Q}{2\pi T}, \quad t > 0,$$

where $a = T/\omega_d$. We can derive the boundary condition at the well by calculating the water flux at time t across a small area element $dA = h(r_0)r_0 d\theta$ along the lateral surface of the cylindrical well, which is a narrow cylinder. This flux is, by Darcy's law,

$$VdA = -Kh_r(r_0,t)h(r_0,t)r_0 d\theta.$$

Therefore the total flux across the lateral surface is

$$-\int_0^{2\pi} Kh_r(r_0,t)h(r_0,t)r_0 d\theta = -Q,$$

or

$$Kh_r(r_0,t)h(r_0,t)r_0 = \frac{Q}{2\pi}.$$

Using the fact that the linearization entails the approximation $Kh(r_0, t) = T$, we obtain the given boundary condition. It is common to reformulate pumping problems in terms of the **drawdown** $s = s(r, t)$ defined by $s = h_0 - h$. Then we have

$$s_t = a \frac{1}{r} \frac{\partial}{\partial r} (r \frac{\partial s}{\partial r}), \quad r > r_0, \quad t > 0,$$

$$s(r, 0) = 0, \quad r > r_0,$$

$$\left(r \frac{\partial s}{\partial r} (r, t) \right)_{r=r_0} = -\frac{Q}{2\pi T}, \quad t > 0.$$

The Theis' approximation involves taking the limit as the well radius r_0 tends to zero. Thus we obtain the problem

$$s_t = a \frac{1}{r} \frac{\partial}{\partial r} (r \frac{\partial s}{\partial r}), \quad r > 0, \quad t > 0,$$

$$s(r, 0) = 0, \quad r > 0,$$

$$\lim_{r \to 0} \left(r \frac{\partial s}{\partial r} (r, t) \right) = -\frac{Q}{2\pi T}, \quad t > 0.$$

This problem can be solved by transforms methods, but here we take a different, more elementary approach. We recall from Chapter 1 that the fundamental solution to the diffusion equation in two dimensions is

$$\frac{1}{4\pi a t} e^{-r^2/4at}.$$

A good strategy would be to try to find the drawdown by superimposing these solutions over time. Therefore, we assume

$$s(r, t) = C_1 \int_0^t \frac{1}{4\pi a \tau} e^{-r^2/4a\tau} d\tau,$$

where the constant C_1 is to be determined from the boundary condition. By direct differentiation of the integral and substitution we obtain

$$r s_r(r, t) = C_1 \int_\infty^{r^2/4at} \frac{1}{2\pi a} e^{-y} dy = -\frac{C_1}{2\pi a} e^{-r^2/4at} = -\frac{Q}{2\pi T}.$$

Taking the limit as $r \to 0$ then yields $C_1 = aQ/T$. Consequently, we have

$$s(r, t) = \frac{Q}{4\pi T} \int_0^t \tau^{-1} e^{-r^2/4a\tau} d\tau,$$

$$= \frac{Q}{4\pi T} \text{Ei}(\frac{r^2}{4at}),$$

where Ei the the **exponential integral** defined by

$$\text{Ei}(z) = \int_z^\infty \frac{1}{\tau} e^{-\tau} d\tau.$$

This expression for the drawdown is **Theis' approximation**. The exponential integral is a special function that is tabulated in standard software packages and tables. For small z we have the approximation [see Abramowitz and Stegun (1962)] $\mathrm{Ei}(z) \simeq -\gamma - \ln z$, where $\gamma \simeq 0.57722$ is Euler's constant. Therefore the drawdown is approximately

$$s(r,t) \simeq -\frac{Q}{4\pi T} \ln(ze^\gamma) = -\frac{Q}{4\pi T} \ln(\frac{r^2 e^\gamma}{4at}), \qquad \frac{r^2}{4at} << 1.$$

This is **Jacob's approximation**. Observe that in a neighborhood of the well, where r is small, we have $s_t \simeq \frac{Q}{4\pi T}t^{-1}$, and thus we obtain very slow variation in the drawdown as $t \to \infty$.

The Theis approximation is a similarity solution. For a further discussion of self-similar flows induced by well-injection in the limit that the well radius goes to zero, see van Duijn and Knabner (1994)

In the transient case, there is a similar diffusion equation for *confined flows* as for unconfined flows. It takes the form

$$S_s h_t = \nabla \cdot (K \nabla h),$$

where S_s is the **specific storage coefficient** having dimensions of inverse length. This coefficient depends upon both the compressibility of the fluid and the compressibility of the fabric, and it is given by $S_s = \rho g(a + \omega\beta)$, where α and β are the fabric and fluid compressibilities, respectively. Another quantitiy of interest is the **storativity** in a confined aquifer, which is defined by $S = bS_s$, where b is the thickness. For comparison of unconfined and confined aquifers, it is generally true that $\omega_d >> hS_s$. Typical values of the parameters might be

$$\omega = 0.3, \quad K = 100 \, \text{m/day}, \quad S_s = 10^{-5} \, \text{m}^{-1}.$$

A detailed discussion of these equations is beyond the scope of this monograph; the references cited at the end of the book can be consulted for information on transient confined flows.

5.3.2 Unsaturated Media

We have said nothing so far about flow in the vadose, or **unsaturated zone**, which is the region above the free surface. In this section, mainly for completeness, we give a brief introduction to a few of the basic ideas in this field, which is often considered a subject in soil physics. An elementary, informative treatment can be found in Fetter (1993).

In the unsaturated zone the pores are not completely filled with water, and there is a mix of water and air. So, in some sense, it is two-phase flow. But we treat only the water phase and assume that the air phase is immobile at atmospheric pressure. The density of the water is constant. To measure the amount of water present we introduce the **volumetric water content** $\theta = \theta(x,t)$, which is the ratio of the volume of water in a small volume element to the total volume of the element. Thus, θ varies between zero and the porosity

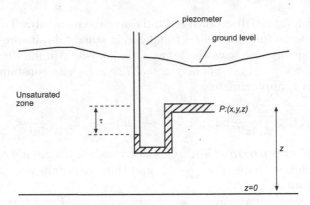

Figure 5.8: Illustration of the suction head τ in the unsaturated zone. The dryness of the zone will pull water from a tube, thereby measuring the suction at a point P.

of the medium. The first experimental observation in unsaturated media is that the conductivity of the medium decreases because of the presence of the air in the pores. Thus, the conductivity is an increasing function of the volumetric water content. The second observation is that the pressure is negative, and therefore the head h is modified by replacing the pressure head by the **suction head** $\tau = \tau(\theta)$. See figure 5.8. Thus, **Darcy's law in unsaturated media** takes the form

$$V = -K(\theta)\nabla h, \quad h = z - \tau(\theta).$$

We can rewrite Darcy's law as

$$V = -K(\theta)\nabla z + K(\theta)\tau'(\theta)\nabla\theta.$$

We next define $D(\theta) = -K(\theta)\tau'(\theta)$, which is the soil–water diffusivity. Then we obtain the velocity in the form

$$V = -K(\theta)\nabla z - D(\theta)\nabla\theta,$$

which is known as the **Buckingham–Darcy law**. This phenomenological law is more complicated than it may first appear because there is generally a hysteresis effect in the suction head—wetting is different from drying. Hence, a graph of τ versus θ looks something like that shown in figure 5.9; the drying and wetting curves are not the same.

The basic governing equation in unsaturated flow is mass balance, and it is known as Richard's equation; it was developed in the early 1930s. To derive the equation we consider an arbitrary unsaturated domain Ω. Then the rate of change of the mass of water in Ω is equal to the net flux of water through the boundary $\partial\Omega$; in symbols,

$$\frac{d}{dt}\int_{\Omega}\rho\theta dx = -\int_{\partial\Omega}\rho V \cdot n dA.$$

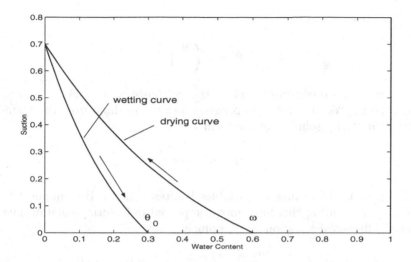

Figure 5.9: The hysteresis effect. Drying a medium from its saturated, intial state $\theta = \omega$ to total dryness, followed by rewetting will lead back to $\theta = \theta_0 < \omega$, the difference $\omega - \theta_0$ being the amount of entrapped air.

Then, applying the divergence theorem in the usual way and appealing to the arbitrariness of the domain, we get, after using the Buckingham-Darcy law,

$$\theta_t = \nabla \cdot (K(\theta)\nabla z + D(\theta)\nabla\theta).$$

This is the higher-dimensional version of **Richard's equation.** In one dimension it becomes

$$\theta_t = (D(\theta)\theta_x)_x, \tag{5.15}$$

which is a nonlinear diffusion equation. If we expand the right-hand side and write it in the form

$$\theta_t = D(\theta)\theta_{xx} + (D'(\theta)\theta_x)\theta_x,$$

then it shows itself as a advection–diffusion equation with diffusion coefficient $D(\theta)$ and nonlinear advection speed $D'(\theta)\theta_x$.

For an example we consider a semi-infinite unsaturated, porous medium occupying $x \geq 0$, initially having water content θ_0. At the inlet boundary we assume a constant "bubble supply" of water content $\theta_1 < \theta_0$. This represents the application of a negative pressure, which corresponds to a positive suction head. Then the initial boundary value problem associated with Richard's equation is

$$\begin{aligned} \theta_t &= (D(\theta)\theta_x)_x, \quad x, t > 0, \\ \theta(x, 0) &= \theta_0, \quad x > 0, \\ \theta(0, t) &= \theta_1, \quad t > 0. \end{aligned}$$

For definiteness, let us take

$$D(\theta) = a \left(\frac{\theta}{\theta_c} \right)^m,$$

where $a, m > 0$ are constants and θ_c is a reference value. For a solution we might expect a "drying front" to progress into the medium from the boundary. We try a similarity solution of the form

$$\theta = f(z), \quad z = \frac{x}{\sqrt{t}}.$$

In this context, the change of variables is often called a Boltzmann transformation. Substituting this form into the partial differential equation gives an ordinary differential equation for f, namely,

$$\frac{a}{\theta_c^m} f^m f'' + \frac{m}{\theta_c^m} f^{m-1}(f')^2 + \frac{z}{2} f' = 0, \quad z > 0,$$

and the initial and boundary conditions become

$$f(0) = \theta_1, \quad f(\infty) = \theta_0. \tag{5.16}$$

To see how to solve this problem we make it simpler by taking $m = 1$ to obtain the differential equation

$$\frac{a}{\theta_c} f f'' + \frac{1}{\theta_c} (f')^2 + \frac{z}{2} f' = 0, \quad z > 0. \tag{5.17}$$

Further simplifying by taking $a = \theta_c = 1$ and then letting $w = f'$, we obtain the first-order system

$$\begin{aligned} f' &= w, \\ w' &= -f^{-1}\left(w^2 + \frac{z}{2} w\right). \end{aligned}$$

At this point we can solve the problem numerically by a shooting method. That is, we use an ODE-solver (e.g., Runge–Kutta) to solve the initial value problem with initial conditions $f(0) = \theta_1$ and $w(0) = \alpha$ for various α until we obtain $f(\infty) = \theta_0$. We request this calculation in the next exercise.

Exercise 78 *Take $f(0) = 0.1$ and $f(\infty) = 0.6$ and use the shooting method outlined in the preceding paragraph to numerically solve the problem (5.16)–(5.17). Plot several time profiles of the drying front.*

5.3.3   The Presence of Solutes

In this paragraph we give a brief glimpse of the next chapter where we study flow and reactions in mineral rocks. If solutes are present in the fluid, then the basic equations for groundwater flow must be expanded to include the solute transport equation. Therefore, we have the system

$$(\rho\omega)_t + \nabla \cdot (\rho V) = \rho q,$$
$$V = -\frac{k}{\mu}\nabla(p + \rho g z),$$
$$(\omega c)_t = \nabla \cdot (\omega D \nabla c) - \nabla \cdot (cV) + r,$$

where c is the concentration of the solute and r is the reaction rate. In nearly all cases we make the tracer hypothesis, which states that the concentration of the solute is so low that the density ρ is constant in the liquid. This assumption simplifies the system and we obtain

$$\omega_t + \nabla \cdot V = q,$$
$$V = -\frac{k}{\mu}\nabla(p + \rho g z),$$
$$(\omega c)_t = \nabla \cdot (\omega D \nabla c) - \nabla \cdot (cV) + r.$$

These five equations have six unknowns: $\omega, c, p,$ and the three components of V. The system accounts for solute transport, porosity changes, and the driving forces that cause the flow. An additional equation is clearly required; we must say something about how porosity changes are related to reaction in the rocks. For example, solutes in the water can react with the mineral substances on the rocks and change the mineral to one of a different volume; this will change the porosity. How these interactions occur is the subject of the next chapter where we discuss reactions and flow in porous rocks.

5.3.4 Thermal Changes

In the next few paragraphs we give a brief discussion of the thermal energy changes that can take place during the evolution of a flow. Generally, we shall not consider these changes here, but it is important to understand what we are leaving out of the models.

Temperature changes can occur in several important and often relevant ways. One of the most obvious energy sources is the generation of heat during chemical reactions that occur in the medium; in the most general cases, the reaction rate will depend on temperature. Other sources of heat might be the heat generated by underground nuclear waste storage repositories or heat produced in deep geological formations by hydrothermal systems (hot springs, magma, etc.). Heat in deep strata, of course, can be a significant source of fluid advection in upper layers, creating vertical ascending flows and even cells of natural rotational flows as in Benard convection. Whenever heat is generated, we must include energy balance as another equation in the governing system.

Heat transfer in porous domains can occur in three obvious ways: conduction in the solid mineral fabric, transport by the fluid, and heat exchange between the solid fabric and the interstitial fluid. The latter is usually removed because we make the assumption that there is associated a *single* temperature $T = T(x,t)$ in the medium that applies to the fabric and fluid alike. Then the only changes in a small volume are caused by transport in the fluid and conduction in fabric. To fix the idea, let Ω be an arbitrary three-dimensional region. If ρ_f, c_f and ρ_s, c_s denote the densities and the specific heats at constant pressure of the fluid and the solid, respectively, then the total amount of heat energy in Ω is given by

$$\text{Total heat energy in } \Omega = \int_\Omega (\omega \rho_f c_f + (1 - \omega)\rho_s c_s)T \, dx.$$

Heat can leave the region by flux through the boundary, which includes both conduction in the fluid and solid, and convection with the fluid. If ndA is an oriented surface element on the boundary of Ω, the heat flux due to conduction is

$$\text{Heat flux due to conduction } = -\omega K_f \nabla T \cdot ndA - (1 - \omega)K_s \nabla T \cdot ndA,$$

where K_f and K_s are the thermal conductivities of the fluid and solid, respectively. The flux through ndA caused by fluid advection is

$$\text{Advective heat flux } = \rho_f c_f TV \cdot ndA.$$

Note that no factor of ω is required in this expression because it contains the filtration velocity V. The net heat flux through the boundary $\partial\Omega$ is then given by

$$-\int_{\partial\Omega} [-\omega K_f \nabla T - (1 - \omega)K_s \nabla T + \rho_f c_f TV] \cdot ndA.$$

If $\overline{H} = \overline{H}(x,t)$ denotes the local heat generation due to sources (e.g., chemical reactions), then thermal energy balance dictates

$$\frac{d}{dt} \int_\Omega (\omega \rho_f c_f + (1 - \omega)\rho_s c_s)T \, dx$$
$$= -\int_{\partial\Omega} [-\omega K_f \nabla T - (1 - \omega)K_s \nabla T + \rho_f c_f TV] \cdot ndA + \int_\Omega \overline{H} \, dx.$$

From the arbitrariness of the domain we get, assuming constant densities, conductivities, and specific heats,

$$(\omega \rho_f c_f + (1 - \omega)\rho_s c_s)T_t = (\omega K_f - (1 - \omega)K_s)\Delta T - \rho_f c_f \nabla \cdot (TV) + \overline{H}$$

where Δ is the Laplacian. If we apply the incompressibility condition and introduce the equivalent heat content and equivalent conductivity defined by

$$\gamma = \frac{\omega \rho_f c_f + (1 - \omega)\rho_s c_s}{\rho_f c_f}, \quad \kappa = \frac{\omega K_f - (1 - \omega)K_s}{\rho_f c_f},$$

then we obtain the thermal energy balance equation

$$\gamma T_t = \kappa \Delta T - V \cdot \nabla T + H \qquad \text{(thermal energy balance)},$$

where $H = \overline{H}/\rho_f c_f$. So, we have obtained an advection-diffusion equation for the temperature T. In one dimension, of course, we have simply

$$\gamma T_t = \kappa T_{xx} - V T_x + H.$$

Typical values of the parameters for various rocks can be found, for example, in de Marsily (1986), p. 281.

In summary, the full equations governing all the effects of solute transport in porous media are quite complicated. They include water balance, species balances, energy balance, reaction kinetics, and constitutive relations like Darcy's law and equations of state relating the various quantities. And there are mechanical–structural changes in the medium, which we have not mentioned. Most models involve some type of simplifying assumptions that permit certain terms in the equations to be ignored. For example, the presumption of isothermal conditions throughout the system allows us to ignore thermal energy balances, and there are many physical situations where this assumption is valid. We must remember, however, consistent with the principles of mathematical modeling, that the deletion of terms and effects in the equations should be the result of a complete dimensional analysis and scaling of the problem where magnitudes of the various terms and scales are determined specifically. We do not undertake such an effort here, but we refer the reader to Phillips (1991) for an thorough discussion of the physical time and spatial scales required to make such simplifying assumptions for a variety of physical problems.

5.4 Reference Notes

Groundwater flow is described in many books at various levels. A classic, advanced reference is Bear (1988); it is often the final word on groundwater analysis, but it is not for the beginner. More elementary, introductory treatments are Verruijt (1970) and Domenico and Schwartz (1990). The standard reference for geologists is Freeze and Cherry (1970). The books by de Marsily (1986) and Hermance (1999) are beginning treatments that have a mathematical flavor. The latter has an outstanding, elementary discussion of the hydraulic head and related ideas. The classic engineering reference is Polubarinova-Kochina (1962).

Chapter 6

Transport and Reactions in Rocks

In this chapter we focus on processes that take place when water flows through a mineral fabric and alters it in some manner. The mineralogy alterations (metamorphism) are caused by an interaction between dissolved particles in water and the solid mineral through which the water is flowing. The interactions can be physical or chemical, and they often lead to significant porosity changes in the domain. One type of interaction is **cementation**, where solid material is deposited at the pore boundaries, and another process is **dissolution**, where solid material is carried away by the liquid. **Replacement reactions** substitute one mineral for another on the surface of the fabric, often with an accompanying change of volume and density. Such processes, which occur on a geologic time scale, lead to the formation of caverns, the deposition of ore deposits and other minerals, and other natural geological structures. Physically, these processes are described in great detail in the hydrogeological literature [for example, see Phillips (1991) or Berner (1971)].

The equations that govern these types of processes are similar to the equations encountered in previous chapters. Generally, there is a mass balance equation that governs the evolution of the dissolved chemical species, a water balance equation (or, continuity equation) that enters in the case of changing volume of liquid, and a chemical kinetics equation that describes the rate(s) that the chemical processes occur. Mathematically, the model equations are interesting because they have a variable porosity appearing in front of the time derivative in the mass balance equation. If the flow is driven by pressure gradients, then Darcy's law must be included. Often, the mixing is caused by thermal convection, initiated by high temperatures at large depths, but we shall not discuss this case.

6.1　Reaction Fronts

In this first section we consider two, simple model reactions that show the variety of wave fronts that can be propagated through mineral rock. This discussion provides some continuity with the earlier discussions of traveling waves and reaction fronts in Chapters 3 and 4, while at the same time raising some crucial issues for flow through porous rocks. The approach here is one of moving from simple examples to a general description of flow–rock interactions.

6.1.1　A Redox Front

Consider a one-dimensional porous medium of constant porosity ω where the flow velocity V is constant. We assume that the water contains a chemical species $\mathbb{C}$ (for example, oxygen) that reacts irreversibly with a mineral $\mathbb{M}$ on the porous rock to produce a product $\mathbb{P}$. Schematically,

$$n\mathbb{C} + l\mathbb{M} \to \mathbb{P},$$

where n and l are the stoichiometric coefficients. So, a chemical-bearing fluid is forced through a mineral-bearing rock with reaction taking place. If the concentrations are small, then the constant porosity assumption is valid. The reaction rate is $R = R(C, M)$, where $C = C(x,t)$ is the concentration of $\mathbb{C}$, measured in moles per volume of water, and $M = M(x,t)$ is the concentration of $\mathbb{M}$, measured in moles per volume of total porous medium. We shall not need to specify the actual form of the reaction rate, only some of its qualitative characteristics. In the standard way, the concentrations C and M satisfy the system of reaction–advection–diffusion equations

$$
\begin{aligned}
C_t &= DC_{xx} - vC_x - n\omega^{-1}R(C,M), \\
M_t &= -lR(C,M),
\end{aligned}
$$

where v is the average velocity. Observe that there are no M-transport terms because $\mathbb{M}$ is immobile.

Intuition dictates that if there is constant concentration M_0 of mineral throughout the medium, then water entering at an inlet boundary far away, carrying $\mathbb{C}$, will cause a redox reaction front to propagate. We expect, therefore to have traveling wave solutions of the form

$$C = C(x - ct), \quad M = M(x - ct)$$

that satisfy the boundary conditions

$$C(-\infty) = C_0, \quad M(-\infty) = 0, \quad C(\infty) = 0, \quad M(\infty) = M_0.$$

Substituting into the governing equations, we obtain

$$
\begin{aligned}
-cC' &= DC'' - vC' - n\omega^{-1}R(C,M), \\
-cM' &= -lR(C,M),
\end{aligned}
$$

where prime denotes the derivative with respect to the moving coordinate $z = x - ct$. Substituting R from the second equation into the first allows the first to be integrated to obtain

$$(v - c)C = DC' - \frac{nc}{l\omega}M + k,$$

where k is a constant of integration. We may determine k by evaluating this expression at $+\infty$; we get $k = (nc/l\omega)M_0$. Then we can evaluate the expression at $-\infty$ to obtain the wave speed

$$c = \frac{l\omega C_0}{l\omega C_0 + nM_0}v.$$

Note that if such a solution exists, then the speed of the reaction front must be less than the average flow velocity, or $c < v$. Moreover, the speed of the redox front is determined only by the stoichiometry and the end states.

Finally, to show that such a solution exists, we can perform a CM-phase plane analysis in much the same way as in the wave front problems in Chapters 3 and 4. The dynamical system is

$$\begin{aligned} DC' &= (v - c)C + \frac{nc}{l\omega}(M - M_0), \\ M' &= nc^{-1}R(C, M). \end{aligned}$$

The end state $(C_0, 0)$ at $z = -\infty$ and the end state $(0, M_0)$ at $z = +\infty$ must be equilibrium states, or $R(C_0, 0) = R(0, M_0) = 0$. These states are clearly critical points of the system and we need to show the existence of a heteroclinic orbit connecting $(C_0, 0)$ to $(0, M_0)$. Generally, if equilibrium is stable, we expect the vector field to point toward the equilibrium curve $R(C, M) = 0$, where the vector field is horizontal. Assuming the equilibrium curve lies strictly above the $C' = 0$ (vertical) nullcline, given by $M = -aC + M_0$, where $a = (l\omega/nc)(v - c) > 0$, we can get the connection. See figure 6.1. Below the vertical nullcline we have $C' > 0$, and above the nullcline we have $C' < 0$. Thus there will be a saddle-node connection representing the redox wave front solution, a schematic of which is shown in figure 6.2. The reader should verify that if the equilibrium curve lies below the vertical nullcline, then the connection goes opposite, connecting $(C_0, 0)$ at $z = +\infty$ and the end state $(0, M_0)$ at $z = -\infty$; so no wave front will exist in this case.

6.1.2 Upstream Fronts

Reaction fronts can also propagate to the left, or upstream, especially when an autocatalytic reaction occurs. An autocatalytic reaction is a simple form of feedback in a chemical system where a chemical aids in its own production. We consider here the model reaction

$$\mathbb{C} + \mathbb{Z} + \mathbb{M} \to (n + 1)\mathbb{Z},$$

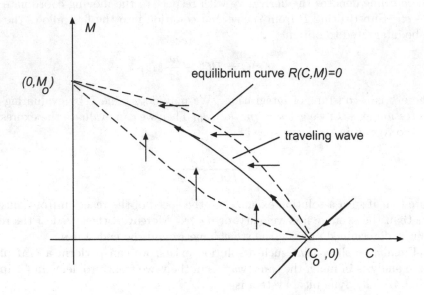

Figure 6.1: CM-phase plane for the redox front.

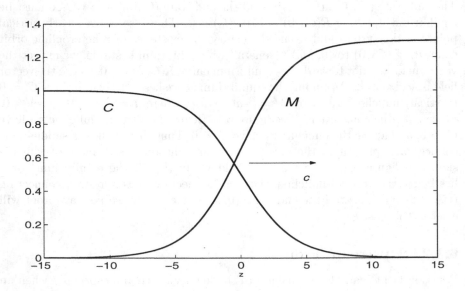

Figure 6.2: Schematic profiles of the solute concentration C and mineral concentration M.

where $\mathbb{C}$ and $\mathbb{Z}$ are mobile species carried by the liquid, and $\mathbb{M}$ is an immobile mineral substance on the porous rock.   Here, n is a stoichiometric amplication factor for the production of $\mathbb{Z}$. The idea is that upstream the fluid is rich in $\mathbb{C}$, with little $\mathbb{Z}$. As before, we assume the average flow velocity $v > 0$ and porosity ω are constants. The reaction rate is given by $R = R(C, M, Z)$, where C, M, and Z denote the concentrations of the given chemical species. We measure the mobile species in moles per volume of liquid and the immobile species by moles per volume of total porous medium. In the usual way the governing equations are

$$\begin{aligned}
C_t &= D_1 C_{xx} - v C_x - \omega^{-1} R(C, M, Z), \\
Z_t &= D_2 Z_{xx} - v Z_x - n\omega^{-1} R(C, M, Z), \\
M_t &= -R(C, M, Z),
\end{aligned}$$

where there are two dispersion constants, one for each of the two mobile species. Now we attempt to find wave front solutions of the form

$$C = C(x - ct), \quad Z = Z(x - ct), \quad M = M(x - ct).$$

For the moment we make no assumptions on the sign of the wave speed c, but we look for solutions where c is negative, which gives left-moving waves. The boundary conditions are suggested by the schematic plots in figure 5.3. Upstream the chemical species $\mathbb{C}$ enters at a constant concentration C_0, while the mineral $\mathbb{M}$ has concentration M_0. The species $\mathbb{Z}$ is in short supply. At $z = +\infty$ the mineral has reacted and the mobile species $\mathbb{Z}$ has been produced via the autocatalytic reaction. Thus, we impose the boundary conditions

$$\begin{aligned}
C &= C_0, \quad M = M_0, \quad Z = 0 \quad \text{as} \quad z \to -\infty, \\
C &= C_\infty, \quad M = 0, \quad Z = Z_\infty \quad \text{as} \quad z \to +\infty.
\end{aligned}$$

Substituting the assumed wave forms into the governing equations gives the system

$$\begin{aligned}
(v - c)C' &= D_1 C'' - \omega^{-1} R, \\
(v - c)Z' &= D_2 Z'' + n\omega^{-1} R, \\
cM' &= R.
\end{aligned}$$

Substituting for R from the last equation into the first two permits the first two to be integrated, and we obtain the two first integrals

$$(v - c)C = D_1 C' - c\omega^{-1} M + k_1, \tag{6.1}$$

$$(v - c)Z = D_2 Z' + nc\omega^{-1} M + k_2, \tag{6.2}$$

where k_1 and k_2 are constants of integration. We determine these constants by evaluating the expressions at $-\infty$, and using the boundary conditions. We get

$$k_1 = (v - c)C_0 + c\omega^{-1} M_0, \quad k_2 = nc\omega^{-1} M_0.$$

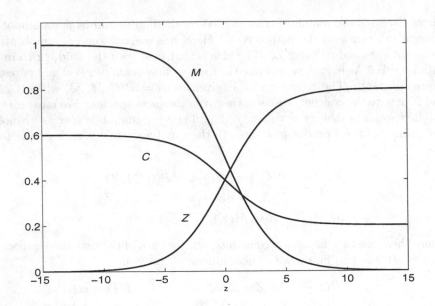

Figure 6.3: Schematic profiles of the mobile concentrations C, Z, and mineral concentration M in a reverse reaction front.

Now evaluating the first integrals at $+\infty$ gives, after some simplification, two expressions for the wave speed, namely,

$$c = \frac{v(C_0 - C_\infty)}{C_0 - C_\infty - \omega^{-1}M_0}$$

and

$$c = \frac{vZ_\infty}{Z_\infty - n\omega^{-1}M_0}.$$

To get $c < 0$ we can take

$$C_0 - \omega^{-1}M_0 < C_\infty,$$

which puts restrictions on the possible values of C_∞. By eliminating c between its two expressions we obtain an equation that then determines Z_∞, that is,

$$Z_\infty = n(C_0 - C_\infty).$$

Consequently, conditions exist under which left-going traveling waves are possible.

To demonstrate the existence of such waves we must pass to a phase plane argument. However, because there are three unknowns, the phase space is three-dimensional and the argument is substantially more difficult. In the case that the two dispersion constants are the same, that is, $D_1 = D_2 = D$, it is possible to reduce the problem to two dimensions. This strategy is a common ploy in the theory of reaction–diffusion equations. Multiplying (6.1) by n and adding it to (6.5) gives the single equation

$$(v - c)(Z + nC) = D(Z + nC)' + k, \quad k = nk_1 + k_2.$$

So, in terms of a new dependent variable $Y = Z + nC$ we have

$$Y' - \frac{v - c}{D}Y = -\frac{k}{D},$$

which is a linear equation with constant coefficients. The only possible bounded solution on $(-\infty, +\infty)$ is $Y = $ constant. Thus $Z + nC$ is constant, or, from the boundary conditions

$$Z + nC = nC_0$$

and this equation permits elimination of Z from the equations. Therefore we have a single, second-order equation for C, namely

$$DC'' + (c - v)C' - \omega^{-1}R(C, M, n(C_0 - c)) = 0, \tag{6.3}$$

where M is given in terms of C and C' by

$$c\omega^{-1}M = k_1 + DC' + (c - v)C.$$

Equation (6.3) can be a convenient starting point for an existence proof in a two dimensional phase plane. We shall not carry out this analysis here, but rather refer the reader to the literature [see Auchmuty *et al* (1984)].

To broaden our perspective, we can imagine two other physical situations where this type of analysis might apply. First, we could consider M to be the concentration of an immobile hydrocarbon that is consumed by a mobile bacterium Z in the presence of a nutrient C. More information on autocatalytic reaction fronts in biological systems can be found in Murray (1993). Second, a similar situation can occur in combustion theory where oxygen-rich gas (C) is forced through a coal-bearing (M) medium. In this case Z plays the role of an "autocatalytic" temperature. Upstream, although oxygen and fuel are present, there is little reaction because of the cool temperatures. A flame front ignited down stream will advance back upstream.

6.2 Porosity-Mineralogy Changes, I

We are now in position to examine the consequences of porosity changes in various model flows in mineral rock. We first analyze a simple model involving moving wave fronts in a one-dimensional porous medium where a chemical solute interacts with the fixed porous fabric to produce an immobile chemical species attached to the matrix. The irreversible reaction is accompanied by a change in porosity. Schematically, we represent the reaction as

$$\mathbb{C} + \mathbb{M} \to \mathbb{S} + \text{products},$$

where $\mathbb{C}$ is the solute, $\mathbb{M}$ is the mineral of the fabric, and $\mathbb{S}$ is the immobile product of the reaction. We assume the flow is pressure driven and is subject to Darcy's law

$$V = -K(\omega)h_x, \qquad (6.4)$$

where $V = V(x,t)$ is the Darcy velocity, $h = h(x,t) = p(x,t)/\rho g$ is the pressure head, $\omega = \omega(x,t)$ is the porosity, and $K = K(\omega)$ is the hydraulic conductivity of the medium. The continuity equation is

$$\omega_t + V_x = 0. \qquad (6.5)$$

If $u = u(x,t)$ denotes the mass concentration of the mobile, chemical solute $\mathbb{C}$, then mass balance requires

$$(\omega u)_t = (D\omega u_x)_x - (Vu)_x - R, \qquad (6.6)$$

where D is the dispersion constant and R is the rate that the solute is consumed by chemical reaction.

The chemical solute is assumed to interact irreversibly with the mineral material forming the porous matrix to produce an immobile chemical species $\mathbb{S}$ whose mass per unit volume of entire porous medium is denoted by $s = s(x,t)$. This product species becomes encrusted on the fixed porous matrix and produces a volumetric change that increases the porosity of the medium. We assume that the porous matrix sites are limited and, as the porosity increases, the number of sites decreases. Thus, the chemical reaction rate is given by

$$R = s_t = uf(\omega), \qquad (6.7)$$

where f is a smooth, nonnegative, decreasing function of ω. Finally, we postulate that the density of the adsorbed species depends on the porosity through a constitutive relation of the form

$$s = g(\omega), \qquad (6.8)$$

where g is smooth, nonnegative and strictly increasing.

Equations (6.4)–(6.8) form the model system for the five unknowns u, ω, s, h, and V on $-\infty < x < +\infty$. It is straightforward to eliminate V and s and reduce the system to

$$\begin{aligned}
g'(\omega)\omega_t &= uf(\omega), & (6.9) \\
(\omega u + g(\omega))_t &= (\omega D u_x)_x + (uK(\omega)h_x)_x, & (6.10) \\
\omega_t &= (K(\omega)h_x)_x. & (6.11)
\end{aligned}$$

We could take, for example, $s = \rho\omega$, and f as $f(\omega) = (\omega_i - \omega)^{2/3}$. Also, as in Chapter 4 on filtration, we could take $\omega = \omega_0 - \beta s$ with kinetics are given by either $s_t = uVF(s)$ or $s_t = uVF(s) - \gamma Vs$.

6.2.1 Dispersive Reaction Fronts

Smooth wave front solutions of (6.9)–(6.11) take the form

$$u = u(\xi), \quad w = w(\xi), \quad h = h(\xi),$$

where $\xi = x - ct$ and c is the unknown wave speed. Here there should be no confusion in using the same lower case letters u, w, and h to denote the wave forms. Substitution of these forms into (6.9)–(6.11) gives

$$-cg'(w)w' = uf(w), \tag{6.12}$$

$$-c(wu + g(w))' = (wDu')' + (uK(w)h')', \tag{6.13}$$

$$-cw' = (K(w)h')', \tag{6.14}$$

where prime denotes $d/d\xi$. We assume the wave front is moving into an equilibrium state with no solute and constant porosity w_0. Accordingly, the boundary conditions at infinity, ahead of the reaction front, are

$$u = 0, \quad w = w_0 \quad \text{at} \quad \xi = +\infty.$$

We assume the inlet, at minus infinity, has a constant solute concentration u_1 and a constant input porosity w_1; furthermore, Darcy's law is assumed to hold. Thus, the boundary conditions at the inlet are

$$u = u_1, \quad w = w_1, \quad -K(w)h' \to v_1 \quad \text{at} \quad \xi = -\infty,$$

where v_1 is the inlet Darcy velocity, and $w_1 > w_0$. The pressure itself will be infinite at $\pm\infty$ because the problem is posed on an infinite interval. Implicit in the assumptions is $f(w_1) = 0$.

We first integrate (6.14) and use the boundary condition at $\xi = -\infty$ to determine the constant of integration. We obtain

$$K(w)h' = c(w_1 - w) - v_1.$$

Substituting this expression into (6.13) and using the boundary condition at $\xi = +\infty$ yields

$$Dwu' = c(g(w_0) - g(w)) + u(v_1 - cw_1).$$

Evaluating this last equation at $\xi = -\infty$ gives the wave speed

$$c = \frac{u_1 v_1}{u_1 w_1 + g(w_1) - g(w_0)} > 0 \tag{6.15}$$

in terms of the limiting equilibrium states. This is a necessary condition for wave front solutions to exist.

Consequently, the porosity and the solute concentration satisfy the two-dimensional dynamical system

$$\frac{dw}{d\xi} = -\frac{uf(w)}{cg'(w)}, \tag{6.16}$$

$$\frac{du}{d\xi} = \frac{1}{Dw}\left(c(g(w_0) - g(w)) + u(v_1 - cw_1)\right). \tag{6.17}$$

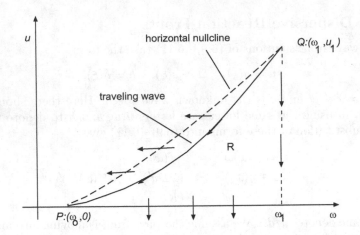

Figure 6.4: Phase portrait for (6.16)–(6.17) showing the unique wave front solution connecting the equilibrium states Q and P.

From the definition of the wave speed, it is easy to check that

$$v_1 - c\omega_1 = u_1^{-1}c(g(\omega_1) - g(\omega_0)) > 0.$$

By the assumptions, the system has critical points in the ωu–phase plane at P: $(\omega_0, 0)$ and Q: (ω_1, u_1). We can show in the usual way that there is a unique heteroclinic orbit connecting Q to P. Refer to figure 6.4. To this end, we note that $\omega' = 0$ along the nullclines $u = 0$ and $\omega = \omega_1$. At the same time, $u' < 0$ along those portions of these nullclines where $\omega_0 < \omega < \omega_1$ and $0 < u < u_1$, respectively. The horizontal nullcline is given by $u' = 0$, or

$$u = \frac{c(g(\omega) - g(\omega_0))}{v_1 - c\omega_1},$$

which is shown in figure 6.4. Easily $\omega' < 0$ on this nullcline. Consequently, the boundary of the region R in figure 6.4 consists entirely of egress or strictly egress points. Finally, one can check the Jacobian (linearized) matrix

$$J(\omega, u) = \frac{\partial(\omega', u')}{\partial(\omega, u)}$$

at the two critical points (e.g., see Walter (1998)). The eigenvalues of the matrix at P are real and opposite sign; thus P is a saddle point. The eigenvalues of the matrix at Q are both positive, and thus Q is an unstable node. Hence there is a unique heteroclinic trajectory connecting the unstable node Q at $\xi = -\infty$ to the saddle P at $\xi = +\infty$. See Hartman (1960). This trajectory represents the reaction front traveling at speed c connecting the two equilibrium states at infinity, and it is indicated in figure 6.4.

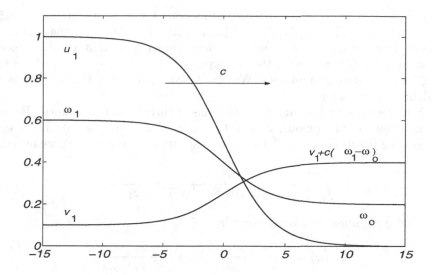

Figure 6.5: Schematic of the wave front solutions for the problem (6.16)–(6.17).

The porosity changes do not depend on the form of the conductivity $K(\omega)$. However, through Darcy's law the pressure must depend on the conductivity. It is straightforward to obtain from the continuity equation that $V(\xi) = c(\omega_1 - \omega(\xi)) + v_1$. Thus, using Darcy's law (6.4),

$$h(\xi) = \int_0^\xi \frac{c(\omega_1 - \omega(\eta)) + v_1}{K(\omega(\eta))} d\eta.$$

The wave forms for u, ω, and s are monotonically decreasing, while the Darcy velocity increases from its value v_1 at $-\infty$ to the value $c(\omega_1 - \omega_0) + v_1$ at $+\infty$. The wave fronts are shown schematically in figure 6.5.

6.2.2   Pure Advection Fronts

Now we examine the simple, purely advective case. In this case smooth wave fronts exist, and we can determine a closed-form solution when the constitutive functions f and g are linear. To this end, if there is no dispersion, then $D = 0$. Thus, smooth traveling wave fronts must satisfy the differential-algebraic system

$$\frac{d\omega}{d\xi} = -\frac{uf(\omega)}{cg'(\omega)},$$
$$u(c\omega_1 - v_1) = c(g(\omega_0) - g(\omega)),$$

where c is the wave speed given in (6.15). This system reduces to the single autonomous equation

$$\frac{d\omega}{d\xi} = -\frac{(g(\omega_0) - g(\omega))f(\omega)}{(c\omega_1 - v_1)g'(\omega)}. \tag{6.18}$$

By the assumptions on the constitutive functions f and g, we note that $d\omega/d\xi < 0$ for $\omega_0 < \omega < \omega_1$, and $d\omega/d\xi = 0$ for $\omega = \omega_0, \omega_1$. Moreover, the right-hand side of the equation is continuous. It follows immediately that a global, smooth solution $\omega = \omega(\xi)$ exits with the properties $\omega \to \omega_1$ as $\xi \to -\infty$ and $\omega \to \omega_0$ as $\xi \to +\infty$. [For example, see Walter (1998)]. Note that the solution is the nullcline shown in figure 6.4.

Now we indicate how analytic solutions of (6.18) can be obtained. Because the equation is autonomous, we can take $\omega(0) = \overline{\omega} \equiv \frac{1}{2}(\omega_1 + \omega_0)$. Then, upon integrating (6.18) from 0 to ξ and changing variables to $r = g(\omega)$, we obtain

$$\int_{g(\overline{\omega})}^{g(\omega)} \frac{dr}{(r - g(\omega_0))(f(g^{-1}(r))} = \frac{\xi}{c\omega_1 - v_1}. \tag{6.19}$$

If f and g are linear functions given by

$$g(\omega) = \frac{\omega - \omega_0}{2(\omega_1 - \omega_0)}, \quad f(\omega) = \frac{\omega_1 - \omega}{2(\omega_1 - \omega_0)},$$

then the integral in (6.19) can be evaluated exactly to obtain, after considerable manipulations,

$$g(\omega) = \frac{1}{2 + 2\beta e^{\gamma \xi}}.$$

The constants are given by

$$\beta = \frac{1 - 2g(\overline{\omega})}{2g(\overline{\omega})} > 0, \quad \gamma = \frac{1}{2(v_1 - c\omega_1)} > 0.$$

Solving the equation for $g(\omega)$ for ω yields

$$\omega(\xi) = \omega_0 + \frac{\omega_1 - \omega_0}{1 + \beta e^{\gamma \xi}}.$$

Then

$$u(\xi) = \frac{\gamma c}{1 + \beta e^{\gamma \xi}},$$

and we have determined exact forms for the wave front solutions.

6.3   Porosity-Mineralogy Changes, II

In this section we advance the analysis of the preceeding section to include more detailed kinetics and specific calculations on a bounded domain.

To reiterate our motivation, when water moves through a porous matrix, a variety of chemical reactions can occur. Among these include reactions that can gradually either enhance or decrease the porosity, and hence the permeability, of the porous fabric. Such processes, which occur on a geologic time scale, are common in nature. As the philosophy of mathematical modeling dictates, a simple mathematical model can often ellucidate the understanding of these processes in a complicated physical system. Models can predict with some degree of accuracy the gross features of the real system, for example, the geologic time that porosity changes occur in a medium. Because the time scale for such changes is so long, laboratory experiments are impossible and one of the only recourses is numerical simulations on model problems. In passing we note that porosity changes can also occur from the deposition of sediments, the stress compaction of the solid fabric, or in other ways [for example, see Domenico and Schwartz (1990), Espedal *et al.* (2000)]. But here our focus in on reaction-porosity relations.

Our model involves a simple replacement reaction of the form

$$\mathbb{C} + N_M \mathbb{M} \rightleftarrows N_S \mathbb{S} + \text{products},$$

where $\mathbb{C}$ is a dissolved chemical species of neglibible volume, $\mathbb{M}$ is an immobile solid mineral attached to the matrix, and $\mathbb{S}$ is an immobile solid product that replaces $\mathbb{M}$ and has a different molecular weight and density. The dissolved products are also assumed to have negligible volume. The dolomitization of limestone is basically this type of reaction, and it can be accompanied by a large (for example, 15%) decrease in volume of the solid phase, and hence an increase in porosity and permeability. We are interested in calculating the response of the model system to various types of initial and boundary conditions and how the porosity responds in time and space. We emphasize again that this study is a mathematical and numerical study of simplified, one-dimensional model equations; the actual, physical system is clearly three-dimensional in nature and the reaction kinetics involve detailed chemistry; moreover, structural and temperature changes, both of which could be significant, are also set aside.

6.3.1 The Model Equations

The governing one-dimensional, partial differential equations for mass balance of the solute, Darcy's law, and the continuity equation are

$$(\omega u)_t = (\overline{\alpha}|V|u_x)_x - (Vu)_x - R, \qquad (6.20)$$

$$V = -\frac{\gamma K(\omega)}{\rho g} p_x, \qquad (6.21)$$

$$\omega_t + V_x = 0, \qquad (6.22)$$

where $u = u(x,t)$ is the concentration (moles per unit volume of liquid) of the solute $\mathbb{C}$, $\omega = \omega(x,t)$ is the porosity, $V = V(x,t)$ is the Darcy velocity, $p = p(x,t)$ is the pressure, and R is the chemical reaction rate for the depletion of $\mathbb{C}$ (in moles per volume of porous medium per unit time). Here $\gamma K(\omega)$ is the hydraulic conductivity, where γ carries the dimensions of velocity and where the dimensionless function $K(\omega)$ depends on the porosity in a way that we shall specify later; ρ and g are the constant density of the fluid and the acceleration due to gravity, respectively. The dispersion coefficient is given by the quantity

$\overline{\alpha}|V|$, where $\overline{\alpha}$ is the dispersivity. Molecular diffusion in the model is neglected, so the flux is only dispersive and advective (the flux is $-\overline{\alpha}|V|u_x + uV$). As we noted, temperature and structural changes can strongly affect volume changes as well, but here these effects are ignored; in this model the changes are strictly reaction-driven, as well as driven by the transport processes that deliver the solute to the mineral surfaces. The overall flow is driven by hydraulic gradients rather than, say, thermal convection.

If $s = s(x, t)$ is the molar concentration per volume of porous medium of the product species $\mathbb{S}$, then

$$s_t = R. \tag{6.23}$$

The reaction rate R depends on both u and ω and has the form

$$R = \overline{k}u_0 f(u/u_0, \omega), \tag{6.24}$$

where $\overline{k}, u_0 > 0$ are constants with units time^{-1}, moles per volume of porous medium, and moles per volume of water, respectively; $\overline{k}$ is the rate constant and $T_r = 1/\overline{k}$ represents a reaction time scale for the problem; f is a dimensionless constitutive function that will be specified in the subsequent discussion. Because temperature variations are not included, the rate constant is temperature-independent. Also, a constitutive relation is required that defines how the porosity changes in terms of the reaction rate and the molecular weights and densities of $\mathbb{S}$ and $\mathbb{M}$. One may take [see, for example, Phillips (1991), p. 127]

$$\omega_t = \beta Q = \beta \overline{k} u_0 f(u/u_0, \omega), \tag{6.25}$$

where β is a constant depending upon the molecular weights and densities having units of volume per unit mole. If $\beta > 0$ then the porosity is enhanced, and if $\beta < 0$, the porosity is decreasing. Specifically, as we show below,

$$\beta = \frac{N_M}{N_S} w_M/\rho_M - w_S/\rho_S, \tag{6.26}$$

where, with the appropriate subscript, w denotes the molecular weight, ρ denotes the density, and N denotes the number of moles in the conversion (the stoichiometric reaction) of the minerals $\mathbb{M}$ and $\mathbb{S}$. Note that the approximate time scale for a porosity change $\Delta\omega$ is

$$\Delta t = \frac{\Delta\omega}{\beta Q}.$$

Researchers have estimated the denominator using the thermal energy balance equation to calculate Q from the Darcy velocity and measured temperature gradients [Wood and Hewitt (1982)].

Because the change in porosity depends linearly on the change in concentration of product species, there is a direct relation between s and ω that takes the form

$$\beta s = \omega - \omega_0, \tag{6.27}$$

where ω_0 is the initial porosity and β is a proportionality constant, which is independent of both x and t, unlike s, ω, and ω_0.

We now derive the last equation from volume considerations. Observe that the volume of the mineral $\mathbb{S}$ at any spatial point x is given by $s(x, t) w_S / \rho_S$ and the volume of the mineral $\mathbb{M}$ is $m(x, t) w_M / \rho_M$. Then the volume of the solid fabric is

$$1 - \omega = s w_S / \rho_S + m w_M / \rho_M. \tag{6.28}$$

Initially, therefore,

$$1 - \omega_0 = s_0 w_S / \rho_S + m_0 w_M / \rho_M.$$

Subtracting these two equations gives

$$\omega - \omega_0 = (s_0 - s) w_S / \rho_S + (m_0 - m) w_M / \rho_M. \tag{6.29}$$

But, by assumption, there is a molar exchange in the reaction (N_M moles of $\mathbb{M}$ produce N_S moles of $\mathbb{S}$), and therefore $N_M(s - s_0) = N_S(m_0 - m)$. That is, molar changes must occur in the same ratio as defined by the stoichiometric equation. Using this to eliminate $(m_0 - m)$ from equation (6.29), we obtain

$$\beta(s - s_0) = \omega - \omega_0, \tag{6.30}$$

with β given by (6.26).

Initial and boundary data must be specified to ensure a well-posed problem. We consider the problem on a bounded spatial domain $0 < x < L$ and $t > 0$, and we assume that initially there is no product species in the porous medium and that the initial porosity is everywhere positive (otherwise flow is blocked). In addition to being phsically reasonable, the positivity assumption prevents a singularity from occuring in equation (6.20). Moreover, we take the initial concentrations and porosity to be

$$s(x, 0) = s_0 = 0, \quad u(x, 0) = u_0(x), \quad \omega(x, 0) = \omega_0(x) > 0, \quad x \in (0, L).$$

Observe that equation (6.27) now follows directly from (6.30).

At the outlet boundary $x = L$ we assume a zero-gradient condition on the concentration u. Because the flux consists of a advective and dispersive part, this assumption does not imply a no-flow boundary; the advective portion of the flux may be nonzero. Thus, we assume

$$u_x(L, t) = 0, \qquad t > 0.$$

We cannot impose conditions on ω, and thus s, at either boundary. At the inlet boundary $x = 0$ we specify the concentration

$$u(0, t) = u_b(t), \qquad t > 0.$$

It is evident that we have a parabolic-type problem in the concentration u, so specifying data at the boundaries is reasonable.

We shall consider one of two types of boundary conditions to drive the flow. We can either impose an overall pressure gradient, i.e.,

$$p(0,t) = p_0, \qquad p(L,t) = 0, \qquad t > 0,$$

or we can impose the Darcy velocity at the inlet,

$$V(0,t) = v_0(t), \qquad t > 0.$$

Recall that flows deep in rocks are often driven by thermal convection, but these are not discussed here.

Next we take up the issue of the form of a model reaction rate R. Clearly, detailed kinetics could be considered [e.g., see Lasaga(1998) for a general discussion of specific chemical reaction rates]. But detailed kinetics are sometimes uncertain in subsurface reactions and rate constants are often unknown up to an order of magnitude, or more; they often depend upon the experimental circumstances. Therefore, in this discussion we argue qualitatively to produce a simple, model, functional form for R that contains a surface area factor, a concentration factor, and a porosity cut-off factor. We first assume, consistent with the law of mass action, that R is proportional to the concentration u of solute brought to the reaction sites at the surface of the solid fabric. Next, the rate should be proportional to the surface area; generally, a rate limiting step is associated with reaction kinetics at the activated sites on the mineral. If we imagine that, at low porosities, the pores of the medium to be composed of cylindrical tubes of fixed radius r, then a unit cross section would have total surface area $n(2\pi r)$, where n is the number of cylinders. The total volume of the pores is $\omega = n(\pi r^2)$, which gives a surface area of $2\sqrt{n\pi}\omega^{1/2}$. On the other hand, at large porosities, we can turn this picture around an imagine the cylinders represent the porous fabric (fibers). In this case the surface area is $2\sqrt{n\pi}(1-\omega)^{1/2}$. Each surface area is itself an appropriate reaction rate factor at one of the extremes ($\omega = 0$ or $\omega = 1$). To accommodate both factors, we define a specific surface area factor given by a combination of these two expressions of the form

$$\text{Surface area factor} \quad = \omega(1-\omega)^{1/2} + (1-\omega)\omega^{1/2}.$$

The first term will be proportional to ω at low porosities and will dominate the second, and the second term will be proportional to $(1-\omega)^{1/2}$ at high porosities and will dominate the first. The proportionality constant is incorporated into the rate constant $\overline{k}$.

Next, we include a final, limiting positive porosity ω_f that cannot be exceeded. Physically, it is clear why this should be so, since the replacement reaction that generates the model amounts to an irreversible replacement of solid mineral by a proportionate amount of product solid.   Note that from (6.27),

$$\omega_f = \omega_0 + \beta s_f,$$

where s_f is the final molar concentration of s.   It is easy to see, using equations (6.27) and (6.29) with $\omega = \omega_f$, that ω_f is related to the final mineral concentration m_f of $\mathbb{M}$ via

$$\omega_f = \omega_0 + \frac{(1 - \omega_0)(1 - r)}{1 + w_S/(\rho_S \beta)},$$

where $r = m_f/m_0$ is the ratio of the final to the initial molar concentrations of $\mathbb{M}$. It is presumed that only a certain amount of the mineral $\mathbb{M}$ can be converted to $\mathbb{S}$ (e.g., surface geometry might limit activation sites). The ratio r should be determined from experimental observation. For example, in some situations it is known that the dolomitization of calcite can result in about a 15% decrease in the solid material.

Accordingly, the function f in the rate law (6.24) incorporates three factors. First, it should have the surface area factor derived above; second, the rate is proportional to u and so we add a normalized factor of u/u_0 to f. Third, the reaction rate should decrease to zero as ω approaches its final value ω_f. Therefore we incorporate a linear factor $\omega - \omega_f$ into f. It may be the case that a nonlinear factor of $\omega - \omega_f$ is more physically appropriate, but a linear factor is used for simplicity. Also, f should always be positive. In the case $\beta > 0$ we have ω_0 smaller than ω_f and the porosity increases, while in the case $\beta < 0$ the porosity decreases from ω_0 to ω_f. To summarize, the function f in (6.25) has the form

$$f = \frac{u}{u_0}|\omega - \omega_f|(\omega(1 - \omega)^{1/2} + (1 - \omega)\omega^{1/2}).$$

Thus, the rate law in this model is based on a geometric factor and not on detailed chemical kinetics. Concerning the latter, the kinetics are not always well-understood, and often the parameters that are input into kinetic models are not known accurately.

It should be noted that for a fixed x equation (6.25) turns into an ordinary differential equation with ω_f behaving like a critical point in an autonomous system, due to the form of f. In particular, ω is either identically ω_f or moves monotonically toward that value.

6.3.2 Dimensionless Equations

Scaled variables can be introduced in the usual way. To reduce the equations to dimensionless form, we choose the length scale L, which is the length of the spatial domain, and the time scale to be the advection time scale $T_c = L/v_0$, where v_0 is a reference velocity scale for V. Pressure is scaled by the pressure $p_0 = \rho g L v_0/\gamma$, and the solute concentration is scaled by a reference concentration u_0, for example, the maximum initial or boundary concentration. The product species s is scaled by $s_0 = \beta^{-1}$ and $v = V/v_0$. To maintain a unit bound on the porosity, we do not scale ω. After scaling, the governing equations

(6.20)–(6.27) in scaled variables take the form

$$(\omega u)_t = (\alpha |v| u_x)_x - (vu)_x - kug(\omega), \tag{6.31}$$

$$v = -K(\omega)p_x, \tag{6.32}$$

$$\omega_t + v_x = 0, \tag{6.33}$$

$$\omega_t = bf(u, \omega), \tag{6.34}$$

$$s = \omega - \omega_0, \tag{6.35}$$

where all of the variables are dimensionless and

$$\alpha = \overline{\alpha}/L, \qquad k = \overline{k}L/v_0, \qquad b = L\beta\overline{k}u_0/v_0 = \beta k u_0 \tag{6.36}$$

are dimensionless quantities. Here, α is the inverse Peclet number, and k is the ratio of the reaction time scale to the advection time scale, that is, $k = T_c/T_r$, and b is a dimensionless time scale for volume changes to occur, which, by the assumptions, is proportional to the reaction time scale. For the numerical calculations, the hydraulic conductivity is assumed to have the Koseny–Carman form [e.g., see p. 62, de Marsily (1986)] for values of ω up to a given value, and constant thereafter. We take

$$K(\omega) = \begin{cases} \frac{\omega^3}{(1-\omega)^2}, & \omega \leq 0.55, \\ 0.8216, & \omega > 0.55. \end{cases}$$

(Another possible form for the hydraulic conductivity is a simple power law.) The rate function g has the form

$$g(\omega) = |\omega_f - \omega|(\omega(1-\omega)^{1/2} + (1-\omega)\omega^{1/2}).$$

The scaled initial data are

$$u(x, 0) = u_0(x), \qquad \omega(x, 0) = \omega_0(x), \qquad x \in (0, 1),$$

and the dimensionless boundary conditions are

$$u(0, t) = u_1(t), \qquad u_x(1, t) = 0, \qquad t > 0,$$

and either

$$p(0, t) = 1, \qquad p(1, t) = 0, \qquad t > 0, \tag{6.37}$$

or

$$v(0, t) = v_1(t), \qquad t > 0, \tag{6.38}$$

where $u_0(x), u_1(t)$, and $v_1(t)$ are dimensionless functions (the original initial and boundary conditions divided by their corresponding scales).

Two cases can be considered, depending upon the desired boundary conditions. If a pressure gradient is driving the flow, then the boundary conditions are given by (6.37). If the flow is driven by velocity, then the boundary condition is given by (6.38).

6.3.3 Pressure-driven flow

We can reduce the number of unknowns in the governing equations by expanding the left-hand side of (6.31), using Darcy's law, the continuity equation, and the rate law. This calculation yields

$$\omega u_t = \alpha(|v|u_x)_x + K(\omega)p_x u_x - kug(\omega). \tag{6.39}$$

Moreover, it follows from the continuity equation and Darcy's law that

$$\omega_t = (K(\omega)p_x)_x, \tag{6.40}$$

and we have the porosity change equation

$$\omega_t = bug(\omega). \tag{6.41}$$

Therefore, there are three equations for u, p, and ω. Here v can be expressed in terms of p and ω by Darcy's law. Now it is clear why the boundary conditions (6.37) are required for pressure, while only initial conditions are needed for ω. In summary, the pressure-driven flow problem can be expressed as a hybrid evolutionary-elliptic system of the form

$$\omega u_t = \alpha(|v|u_x)_x + K(\omega)p_x u_x - kug(\omega), \tag{6.42}$$
$$\omega_t = bug(\omega), \tag{6.43}$$
$$(K(\omega)p_x)_x = bug(\omega), \tag{6.44}$$

with initial and boundary conditions

$$\begin{aligned}
u(x,0) &= u_0(x), & w(x,0) = w_0(x), & \quad x \in (0,1), \\
u(0,t) &= u_1(t), & u_x(1,t) = 0, & \quad t > 0, \\
p(0,t) &= 1, & p(1,t) = 0, & \quad t > 0.
\end{aligned}$$

In Section 6.4.5 we indicate how this problem can be solved numerically.

6.3.4 Velocity-Driven Flow

When a forced filtration velocity at the inlet boundary drives the flow, the governing equations can be written as a nonlocal system of integrodifferential equations in terms of just two unknowns, u and ω. To this end note that

$$-v_x = bug(\omega).$$

Integration gives

$$v(x,t) = -\int_0^x bu(y,t)g(\omega(y,t))dy + v(0,t). \tag{6.45}$$

When this expression is substituted into the equation

$$\omega u_t = \alpha(|v|u_x)_x - vu_x - kug(\omega). \tag{6.46}$$

for v, there results a single equation in terms of ω and u. Next, equation (6.41) can serve to eliminate the term p_x from (6.40), yielding the desired system. Now it is clear that boundary conditions on u and the boundary velocity $v(0, t)$, along with the initial data for ω and u, are sufficient to determine the solution. In summary, the pressure-driven flow problem can be formulated as a two-dimensional evolutionary system with a non-local term of the form

$$
\begin{aligned}
\omega u_t &= \alpha(|v|u_x)_x - vu_x - kug(\omega), \\
\omega_t &= bug(\omega), \\
u(x, 0) &= u_0(x), \quad \omega(x, 0) = \omega_0(x), \quad x \in (0, 1), \\
u(0, t) &= u_1(t), \quad u_x(1, t) = 0, \quad t > 0, \\
v(0, t) &= v_1(t), \quad t > 0.
\end{aligned}
$$

where $v(x, t)$ is specified in terms of ω and u by the formula (6.45).

It is easy to see that the two initial-boundary-value problems discussed above are essentially equivalent. Suppose first that u and ω are solutions to the pressure-driven problem stated above. Numerical quadrature can determine the input $v(0, t)$. To see this, fix t and define

$$
\phi(x, t) \equiv v(x, t) - v(0, t) = -\int_0^x bu(y, t)g(\omega(y, t))dy. \tag{6.47}
$$

For a given u and ω, the function $\phi(x, t)$ is *known*. On the other hand, the boundary conditions and Darcy's law imply that

$$
\begin{aligned}
1 &= p(0, t) - p(1, t) = -\int_0^1 p_x dx = \int_0^1 \frac{v(x, t)}{K(\omega(x, t))} dx \\
&= v(0, t) \int_0^1 \frac{dx}{K(\omega(x, t))} + \int_0^1 \frac{\phi(x, t)}{K(\omega(x, t))} dx.
\end{aligned}
$$

This equation then determines $v(0, t) = v_1(t)$, and it follows that u and ω are solutions to the velocity-driven problem.

Conversely, if u and ω are solutions to the velocity-driven problem, then the function $\phi(x, t)$ is determined, as well as the velocity function $v(x, t)$ by formula (6.47). The pressure can be defined by requiring that it obey Darcy's law (6.32), which implies the quadrature formula

$$
p(x, t) \equiv p_0(t) - \int_0^x \frac{v(y, t)}{K(\omega(y, t))} dy,
$$

where $p_0(t)$ is an arbitrary function of t. It follows that u and ω are solutions to the pressure-driven problem with one modification: the boundary pressure values are no longer 1 and 0 and indeed, the pressure function itself is determined only up to an arbitrary function of time. This is not surprising in view of the fact that the only way in which p appears in the original dimensionless equations (6.31)–(6.35) is by way of the spatial derivative p_x, knowledge of which can only determine p up to a time varying constant of integration.

6.3.5 Numerical Results

Numerical calculations on the preceding problems were performed in MATLAB version 5 (1988). The pressure-driven problem (6.42)–(6.44) can be solved numerically by a straightforward, implicit, finite-difference scheme (see Appendix A). The idea is to use (6.43) to advance ω to the first time step from the initial data in ω and u. Then use (6.44) to solve, implicitly, a boundary value problem for pressure p at the first time step. Finally, one can use (6.42) to implicitly solve the boundary value problem for the concentration u at the first time step. Then one continues the process repeatedly, thereby marching forward in time. At each time step the two boundary value problems for p and u are linear and can be solved by a Crank–Nicolson-type formula (Appendix A). The velocity driven problem was solved by the method of lines (MOL) (Appendix A). Because this problem is purely evolutionary in time (albeit complicated by the non-local term), it is possible to develop a MOL code to use built-in ODE solvers in MATLAB (along with some numerical quadrature routines to recover v and p.). Owing to the relative stiffness of the problem, a finite-difference scheme is much slower because of the small time steps required for stability. With the MOL code, the MATLAB **ode15s** routine can handle the relative stiffness of the problem fairly well.

One difficulty in performing the calculations is that there are wide ranges for the various parameters in the problem. Reaction rate constants can vary over four orders of magnitude; a value of $\overline{k} = 10^{-10}$ per second corresponds to a slow reaction, and a value of $k = 10^{-6}$ per second corresponds to a fast reaction. Length scales can vary over $10^2 < L < 10^5$ meters, and filtration velocities can vary over $10^{-8} < v_0 < 10^{-6}$ meters per second, the low end representing a few meters per year. Dispersivities $\overline{\alpha}$ have been measured around a few centimeters in laboratory sand columns and up to a hundred meters in field experiments in fractured rock. Thus, $0.01 < \overline{\alpha} < 100$ meters. The factor β is on the order of 10^{-6} cubic meters per mole; for example, the value of $\beta = 8.01(10)^{-6} \text{m}^3/\text{mole}$ is computed from the assumed reaction $\mathbb{C} + 2\mathbb{M} \rightleftarrows \mathbb{S}$ + products, where two moles of $\mathbb{M}$ (calcium carbonate: density $\rho_M = 2.71 \text{ gm/cm}^3$ and atomic weight $w_M = 100.09$) convert into one mole of $\mathbb{S}$ (dolomite: density $\rho_S = 2.80 \text{ gm/cm}^3$ and atomic weight $w_S = 184.4$). The actual reaction is

$$2\text{CaCO}_3 + \text{Mg}^{2+} \rightarrow \text{CaMg(CO}_3)_2 + \text{Ca}^{2+}.$$

Consequently, the dimensionless constants in the problem can have tremendous variation:

$$10^{-3} < k < 10^7, \quad 10^{-7} < \alpha < 1, \quad 10^{-8} u_0 < b < 10^2 u_0.$$

For numerical calculations some choices of these parameters can lead to difficulty because the problem becomes stiff. Their choice, of course, depends upon the system studied.

For the calculations illustrated here, the values of the dimensioned parameters, in mks units, are given in the following table; the porosity is typical of limestone and secondary dolomites; the Darcy velocity corresponds to 2 m/yr.

Parameter	Value
ω_0	0.2
L	1000 m
$\overline{\alpha}$	100 m
v_0	$3.17(10)^{-8}$m/s
$\overline{k}$	$1.0(10)^{-9}$/s
u_0	1 mole/m^3
β	$8.01(10)^{-6}$m^3/mole
ω_f	0.44

Values of the dimensionless parameters are computed from the dimensionless parameters by the formulas in equation (6.36).

Parameter	α	b	k
Value	0.1	0.000252	31.5

The advection time scale for these parameter choices is 500 years, and for a 15% porosity change (corresponding to the given value of ω_f) to occur, it takes about $0.15/0.00091 \cong 165$ time steps, or about 82,400 years, to see these discernible changes. The entire medium undergoes nearly complete metamorphosis in about 540,000 time steps, or 270 million years. Using a MOL code on a desktop PC, a calculation with 41 spatial nodes can be completed, along with graphics using a programmed interface, in a few minutes. Figure 6.6 shows the molar concentration surface u after 150,000 time steps, or about 75 million years. Similarly, figures 6.7 and 6.8 show the porosity and pressure surfaces, respectively. Velocity profiles are nearly constant across the medium, increasing about three times the initial value. The concentration and porosity changes resemble traveling wave fronts moving into the medium.

For closely related work on various mathematical issues associated with porosity changes, and for additional references, the reader can consult Grindrod (1993, 1996), Cohn *et al.* (2000), Logan (1999), and Ledder and Logan (2000). Some of the content of this section is taken from Logan, Petersen, and Shores (2001).

6.4   General Equations of Reaction and Flow

Having worked through some specific examples, we now turn to a general, macroscopic, phenomenological description of transport and reactions in porous mineral rock in three dimensions. Basically, there are evolution equations for the minerals and for the solutes, and there are dynamics that describe the chemical processes.

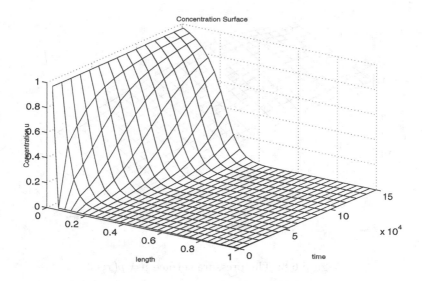

Figure 6.6: Time snapshots of the scaled solute concentration $u = u(x, t)$.

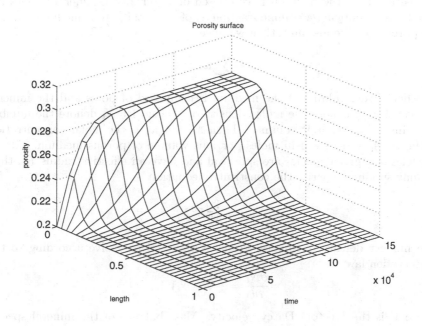

Figure 6.7: Time snapshots of the porosity $\omega = \omega(x, t)$.

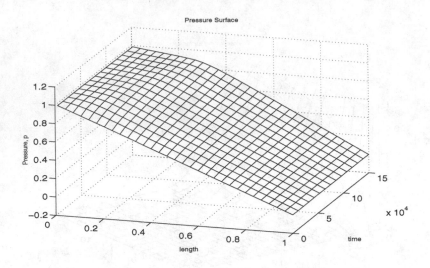

Figure 6.8: The pressure surface $p = p(x, t)$.

6.4.1   Mineral-Porosity Changes

We assume that the medium is composed of water and I mineral species $\mathbb{M}_i$ $(i = 1, \dots I)$ occupying a volume fraction ω_i of the total porous medium. If ω is the porosity of the medium, then we have

$$\omega + \sum \omega_i = 1.$$

In other words, volume of the medium is composed of pores and the mineral grains. Let ρ_i denote the molar density of $\mathbb{M}_i$ and let N_i denote the number of grains of $\mathbb{M}_i$ per unit volume. Then $M_i = \rho_i \omega_i$ is the molar concentration of the ith species, given in moles per unit volume of porous medium. If the particles of a given species are spherical and have an effective radius R_i, then we may specifically write the constitutive equation

$$\omega_i = \frac{4}{3} \pi R_i^3 N_i.$$

The number density of mineral grains is assumed to evolve according to the conservation law

$$\frac{\partial N_i}{\partial t} = -\nabla \cdot (N_i V),$$

where V is the (vector) Darcy velocity. Mass balance of the mineral species gives, assuming only advection and reaction,

$$\frac{\partial M_i}{\partial t} = -\nabla \cdot (M_i V) + \widetilde{Q}_i,$$

where $\widetilde{Q}_i$ is the rate of formation of the ith component in moles per unit volume-time. We assume that the molar densities are constant, and so we can write

$$\rho_i \frac{\partial \omega_i}{\partial t} = -\rho_i \nabla \cdot (\omega_i V) + \widetilde{Q}_i.$$

It easily follows that the effective radii satisfy the equations

$$\frac{\partial}{\partial t}\left(\frac{4}{3}\pi R_i^3\right) = -V \cdot \nabla \left(\frac{4}{3}\pi R_i^3\right) + \frac{1}{\rho_i N_i}\widetilde{Q}_i.$$

If the mineral species are immobile, then there is no advective term and

$$\frac{\partial}{\partial t}\left(\frac{4}{3}\pi R_i^3\right) = \frac{1}{\rho_i N_i}\widetilde{Q}_i.$$

This equation describes how the radius of the grains evolve in terms of the reaction rate.

6.4.2   Solute Changes

Next we assume that there are J aqueous species $\mathbb{C}_j$ $(j = 1, ..., J)$ having molar concentrations C_j. Mass balance for each of of these species implies

$$\frac{\partial}{\partial t}(\omega C_j) = -\nabla \cdot \phi_j + Q_j,$$

where ϕ_j is the vector flux of the jth species in moles per area-time and Q_j is the rate of formation of that species. In the simplest case, the flux will have the form

$$\phi_j = -\omega_j D_j \nabla C_j + C_j V,$$

which consists of a dispersive and an advective part.

6.4.3   Reaction Kinetics

The reaction rates are constitutive equations given by the law of mass action. Generally, there are P reactions of the form

$$\sum_{j=1}^{J} \lambda_{pj}\mathbb{C}_j + \sum_{i=1}^{I} \widetilde{\lambda}_{pi}\mathbb{M}_i = 0, \quad p = 1, ..., P,$$

where the λ_{pj} and the $\widetilde{\lambda}_{pi}$ are the stoichiometric coefficients; for example, λ_{pj} represents the net number of molecules of $\mathbb{C}_j$ produced in the pth reaction. Then

$$\widetilde{Q}_i = \sum_{p=1}^{P} \widetilde{\lambda}_{pi}W_p, \quad Q_j = \sum_{p=1}^{P} \lambda_{pj}W_p,$$

where W_p is the rate of the pth reaction. Mass action specifies that the rate for the pth reation is the product of the concentrations of the reactants entering the reaction, each to the power of the number of moles of that component. The constant of proportionality, called the rate constant, often depends on temperature, but this analysis does not include thermal energy changes.

In summary, the governing equations for mineral mass balance and solute mass balance are

$$\frac{\partial M_i}{\partial t} \;=\; -\nabla \cdot (M_i V) + \sum_{p=1}^{P} \tilde{\lambda}_{pi} W_p,$$

$$\frac{\partial}{\partial t}(\omega C_j) \;=\; -\nabla \cdot \phi_j + \sum_{p=1}^{P} \lambda_{pj} W_p.$$

Example 79 *The simple, single reaction*

$$2\mathbb{M}_1 + 4\mathbb{C}_1 + \mathbb{C}_2 \to \mathbb{C}_1 + \mathbb{M}_2 \quad or \quad -2\mathbb{M}_1 + \mathbb{M}_2 - 3\mathbb{C}_1 - \mathbb{C}_2 = 0,$$

has rate $W = kM_1^2 C_1^4 C_2$, *where* k *is the rate constant.   The stoichiometric coefficients are* $-2, +1, -3, -1$. *The rates of consumption of* $\mathbb{M}_1, \mathbb{M}_2, \mathbb{C}_1,$ *and* $\mathbb{C}_2$ *are*

$$\widetilde{Q_1} = -2W, \quad \widetilde{Q_2} = W, \quad Q_1 = -3W, \quad Q_2 = -W.$$

(Note, if the reaction were reversible, that is,

$$2\mathbb{M}_1 + 3\mathbb{C}_1 + \mathbb{C}_2 \rightleftharpoons \mathbb{C}_1 + \mathbb{M}_2,$$

then the rate is $W = k_f M_1^2 C_1^4 C_2 - k_b C_1 M_2$, *where* k_f *and* k_b *are the forward and backward rate constants, respectively.)   The governing equations are*

$$\frac{\partial M_1}{\partial t} \;=\; -\nabla \cdot (M_1 V) - 2W,$$

$$\frac{\partial M_2}{\partial t} \;=\; -\nabla \cdot (M_2 V) + W,$$

$$\frac{\partial}{\partial t}(\omega C_1) \;=\; -\nabla \cdot \phi_1 - 3W,$$

$$\frac{\partial}{\partial t}(\omega C_2) \;=\; -\nabla \cdot \phi_2 - W.$$

Example 80 *In some reactions, because of the abundance of one of the reactants, the concentration of that reactant does not change significantly during the course of reaction.   In this case the concentration of that substance is taken to be constant.   For example, suppose we have the reaction*

$$2\mathbb{C} + \mathbb{M} \to inert\ products.$$

By the law of mass action the reaction rate is $W = kMC^2$. *However, compared to the solute* $\mathbb{C}$, *if the mineral* $\mathbb{M}$ *is abundant and is not consumed to any* *large degree in the reaction, then we can assume M is essentially constant and incorporate its factor into the rate constant k. In this case, the reaction rate is usually written $W = kC^2$.*

Example 81 *(The quasi-steady-state hypothesis) In some reactions short-lived intermediates may be so reactive that they do not accumulate in large quantities and so they are difficult to detect. Their presence, however, may be essential in the overall reaction mechanism. For example, consider the reaction*

$$\mathbb{C}_1 \;\rightleftharpoons\; \mathbb{C}_2,$$
$$\mathbb{C}_2 \;\rightarrow\; \mathbb{C}_3,$$

where $\mathbb{C}_2$ is such an intermediary. The reaction rate equations are

$$\frac{dC_1}{dt} = -k_f C_1 + k_b C_2,$$
$$\frac{dC_2}{dt} = k_f C_1 - k_b C_2 - k C_2,$$
$$\frac{dC_3}{dt} = k C_2.$$

*In this case we assume that $\frac{dC_2}{dt} = 0$, i.e., $C_2 = $ constant, which is the **quasi-steady-state hypothesis** (QSSH). Thus, C_2 is so highly reactive that when it is formed, it rapidly reverts back to $\mathbb{C}_1$ or decomposes into $\mathbb{C}_3$. The QSSH assumes an equilibrium concentration of $\mathbb{C}_2$ is formed early in the reaction and changes only slowly thereafter; that is, the rate of its formation equals the rate of its consumption. This assumption would force $k_f C_1 - k_b C_2 - k C_2 = 0$, or*

$$C_2 = \frac{k_f C_1}{k_b + k}.$$

This hypothesis then dictates the form of the rate law, namely

$$\frac{dC_1}{dt} = -k_f C_1 + k_b \frac{k_f C_1}{k_b + k} \equiv -K C_1.$$

Therefore $C_1 = const.e^{-Kt}$. On many occasions the QSSH leads to correct predictions where singular perturbation methods fail.

6.5 Reference Notes

Much of the preceeding description of the flow and reactions in porous rock lends itself to an analysis of pattern formation in geological structures. Just as reaction-diffusion equations in biological systems [e.g., see Murray (1993)], through the pioneering work of A. Turning, led to answers to such questions as "Why leopards have spots"?, so it is with reaction–advection–dispersion equations in the geosciences. The study of pattern formation in this context is called *geochemical self-organization.* Using the macroscopic, nonequilibrium model discussed above, many researchers have shown how a variety of patterns can form through the interaction of transport and chemistry in porous structures, instabilities, and nonlinear dynamics.   One example is the genesis of banded patterns of precipitants called Liesegang bands that occur when coprecipitant ions interdiffuse and lock-in a permanent pattern. It is beyond the scope of this work to analyze these processes, and thus we refer the reader to the literature. An excellent starting point is Ortoleva (1994).

Appendicies

Appendices

A The Finite-Difference Method

Most real-world problems faced in industry or practical, applied science lead to PDE models that are too complicated to solve analytically. Nonlinearities, complicated boundaries, and heterogeneous media all contribute to difficulties. Therefore, PDEs are almost always solved numerically, on a computer. Even if a PDE problem can be solved analytically, usually the solution is in the form of a difficult integral or an infinite series, thereby forcing a computer calculation anyway. By a numerical or computational method we mean something different from asking a computer algebra program (like Maple or Mathematica) to return a formula, which is possible for some classes of equations. By a numerical method we generally mean that the continuous PDE model is replaced by a discrete model that can be solved on a computer in finitely many steps. The result is a discrete solution where the solution surface is known approximately at only finitely many points.

Among the most common numerical techniques for solving PDEs are the finite-difference method (FDM), the finite-element method (FEM), characteristic methods, and the method of lines (MOL). There are other methods as well, like the boundary integral method, Monte Carlo methods, and so forth. The entire area of numerical PDEs is one of the most active in science and new algorithms, as well as modifications of older ones, appear regularly in the literature. The FDM is the simplest of all the methods to program, and it is the easiest to understand. However, there are some difficulties. Explicit FDMs may require a very small time step to maintain stability, resulting in a large number of calculations; and for problems where advection dominates dispersion (large Peclet number flows), steep, propagating fronts may not be approximated well. The class of FEMs is harder to program and conceptually more difficult, but it is better adaptable to unusual domains. On the whole, FEMs are more flexible than FDMs and generally more superior. Characteristic methods are particularly suitable for hyperbolic problems involving wave propagation, and they can be used along with particle tracking methods to solve advection-dominated flows with FDMs. The MOL is a hybrid FDM that adapts itself to the powerful software packages that solve systems of ordinary differential equations.

In these appendices we focus attention on the finite-difference method for the advection–dispersion–adsorption equation and on the method of lines. The discussion is a tutorial rather than a careful analysis, and sample programs are listed. The analysis of numerical methods, scientific computation, and numerical modeling are among the most active areas of research in applied science as investigators seek to find faster, more accurate algorithms. The reader will find a large amount of literature on the subject, and we refer the reader to the many books on numerical computation for a detailed treatment.

Explicit Method

The idea of the finite-difference method is to replace the partial derivatives in the PDE by difference quotient approximations and then let the computer solve the resulting difference equation. One should be familiar with this strategy for ordinary differential equations. For review, we quickly illustrate the Euler method for numerically solving the initial value problem

$$y' = f(t, y), \quad 0 < t < 1; \quad y(0) = y_0.$$

We discretize the interval $0 \le t \le 1$ by defining a finite number of discrete, or nodal, points $t_n = n\Delta t$, $n = 0, 1, \ldots, N$, where Δt is the *stepsize* given by $\Delta t = 1/N$. At each point t_n we wish to determine a value Y_n that is a good approximation to the exact value $y(t_n)$ of the solution. From differential calculus we know that $y'(t_n)$ can be approximated by a forward difference quotient

$$y'(t_n) \approx \frac{y(t_{n+1}) - y(t_n)}{\Delta t},$$

provided Δt is small. Consequently, we approximate the differential equation by

$$\frac{Y_{n+1} - Y_n}{\Delta t} = f(t_n, Y_n), \quad n = 0, 1, \ldots, N,$$

or

$$Y_{n+1} = Y_n + \Delta t f(t_n, Y_n), \quad n = 0, 1, \ldots, N.$$

Now, given $Y_0 = y_0$, this difference equation provides a numerical algorithm to *explicitly* calculate the approximations $Y_1, Y_2, \ldots$, recursively. This same strategy, the idea of marching forward in time, goes over to evolution problems in PDEs.

We will illustrate an explicit FDM for an initial boundary value problem governed by the advection–dispersion–adsorption equation with Dirichlet boundary conditions:

$$u_t = Du_{xx} - vu_x + F(u), \quad 0 < x < L, \quad t > 0, \tag{A.1}$$

$$u(0, t) = g(t), \quad u(L, t) = h(t), \quad t > 0, \tag{A.2}$$

$$u(x, 0) = f(x), \quad 0 < x < L. \tag{A.3}$$

The first step to discretize the region of spacetime where we want to obtain a solution. In this case the region is $0 \le x \le L$, $0 \le t \le T$. We have put a bound on time because in practice we only solve a problem up until a finite time. Discretizing means defining a lattice of points in this spacetime region by

$$x_j = j\Delta x, \quad t_n = n\Delta t, \quad j = 0, 1, \ldots, J; \quad n = 0, 1, \ldots, N,$$

where the fixed numbers Δx and Δt are the spatial and temporal stepsizes, respectively. Here, $\Delta x = L/J$ and $\Delta t = T/N$. The integer J is the number of subintervals in $0 \leq x \leq L$ and N is the number of time steps to be taken. Figure A1 shows the lattice of points (the lattice points are also called grid points or nodes). At each node (x_j, t_n) of the lattice we seek an approximation, which we call U_j^n, to the exact value $u(x_j, t_n)$ of the solution. Note that the superscript n refers to time and the subscript j refers to space. We can regard U_j^n as a two-dimensional array, or matrix, where n is a row index and j is a column index. To obtain equations for the U_j^n we replace the partial derivatives in the PDE by their difference approximations. The time derivative is approximated by a forward difference

$$u_t(x_j, t_n) \approx \frac{u(x_j, t_{n+1}) - u(x_j, t_n)}{\Delta t},$$

and the first spatial partial derivative is approximated by a centered difference

$$u_x(x_j, t_n) \approx \frac{u(x_{j+1}, t_n) - u(x_{j-1}, t_n)}{2\Delta x}.$$

The second derivative is approximated by

$$u_{xx}(x_j, t_n) \approx \frac{u(x_{j-1}, t_n) - 2u(x_j, t_n) + u(x_{j+1}, t_n)}{\Delta x^2}.$$

These formulas are the usual forward difference approximation for a first derivative, and the second is a difference approximation for a second derivative; the latter follows from a Taylor approximation. So the PDE (A.1) at the point (x_j, t_n) is replaced by the difference equation

$$\frac{U_j^{n+1} - U_j^n}{\Delta t} = D\frac{U_{j-1}^n - 2U_j^n + U_{j+1}^n}{\Delta x^2} - v\frac{U_{j+1}^n - U_{j-1}^n}{2\Delta x} + F(U_j^n), \qquad \text{(A.4)}$$

or, upon solving for U_j^{n+1},

$$U_j^{n+1} = (r+s)U_{j-1}^n + (1-2r)U_j^n + (r-s)U_{j+1}^n + \Delta t F(U_j^n), \qquad \text{(A.5)}$$

where

$$r = \frac{D\Delta t}{\Delta x^2}, \quad s = \frac{v\Delta t}{2\Delta x}. \qquad \text{(A.6)}$$

Observe that this equation relates the approximate values of the solution at the four points $(x_{j-1}, t_n), (x_j, t_n), (x_{j-1}, t_n), (x_j, t_{t+1})$. These four points form the *computational atom* for the difference scheme (see figure 1).

The difference equation (A.5) gives the approximate solution at the node (x_j, t_{t+1}) in terms of approximations at three earlier nodes $(x_{j-1}, t_n), (x_j, t_n),$

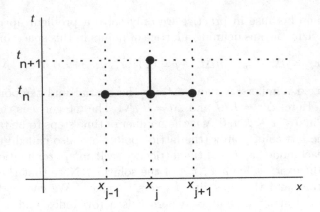

Figure 1: The discrete lattice and the computational atom for the adsorption–advection–dispersion equation.

and (x_{j-1}, t_n). Now we see how to fill up the lattice with approximate values. We know the values at $t = 0$ from the initial condition (A.3). That is we know

$$U_j^0 = f(x_j), \quad j = 0, 1, \dots, J. \qquad (A.7)$$

From the boundary conditions (A.2) we also know

$$U_0^n = g(n\Delta t), \quad U_J^n = h(n\Delta t), \quad n = 1, 2, \dots, N. \qquad (A.8)$$

The difference formula (A.5) can now be applied at all the interior lattice points, using the values at the $t = 0$ level to compute the values at $t = t_1$ level, then using those to compute the values at the $t = t_2$ level, and so on. Therefore, using the difference equation we can march forward in time, continually updating the previous temperature profile. We can think of the process as filling out the the array U_j^n, row by row.

A finite-difference scheme, or algorithm, like (A.5) is called an *explicit scheme* because it permits the explicit calculation of an approximation at the next time step in terms of values at a previous time step. Because of the truncation error that is present in (A.4) caused by replacing derivatives by finite differences, the scheme is not always accurate. If the time step Δt is too large, then highly inaccurate results will be obtained. It can be shown that we must have the *stability condition*

$$r = \frac{D\Delta t}{\Delta x^2} \le \frac{1}{2} \qquad (A.9)$$

for the scheme to converge.

The following MATLAB program solves the boundary value problem in (A1)–(A3) with homogeneous Dirichlet boundary conditions, $D = 0$, $v = 0.2$, and $F(u) = 0$. The solution, represented as concentration profiles, is shown in figure 2. Observe that we have solved a purely advection problem since $D = 0$.

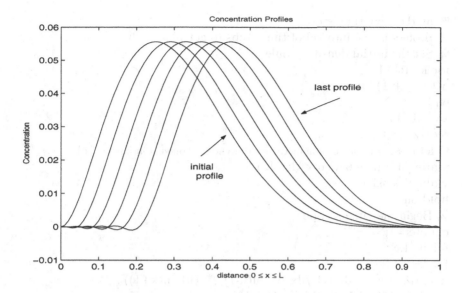

Figure 2: Purely advective concentration surface when $f(x) = 5x^2(1-x)^5$.

The oscillations near $x = 0$, which are a numerical artifact, are common in advection-diffusion problems solved by explicit FDMs. Here, the final profile is at time $T = Ndt = (1000)(.001) = 1.0$. The profiles occur every 200 time steps.

MATLAB Program (Explicit FDM)
function reactdiff
% Description: 1D explicit code for an adsorption-advection-dispersion equation
% u_t=Du_xx-vu_x+Q(u)
% with data u(x,0)=f(x), au(0,t)+bu_x(0,t)=g(t), cu(L,t)+du_x(L,t)=h(t)
% No false boundary introduced
L=1; D=0; v=0.2;
a=1; b=0; c=1; d=0;
J=100; dx=L/J;
dt=0.001; N=1000; profiles=200;
r=D*dt/(dx^2);
if r>0.5
var='Unstable–Take a smaller time step!'
break
end
% L is the interval length, D is the diffusivity, v is the velocity.
% a,b,c,d are constants that occur in the boundary conditions.
% N is the number of time steps and J is the number of spatial steps.
% u is the density at the current time and unext is the density

```
% at the next time step.
% profiles is the interval of time steps to get plots
% Set the initial density profile
for k=1:J+1
u(k)=f((k-1)*dx);
end
z=[0:L/J:L];
plot(z,u,'r')
title('DENSITY PROFILES FOR REACTION-CONVECTION-DIFFUSION EQUATION')
xlabel('distance 0 \leq x \leq L')
ylabel('Density')
hold on
% Begin the time stepping
counter=1;
for n=1:N
for k=2:J
unext(k)=u(k)+dt*((D/dx^2)*(u(k-1)-2*u(k)+u(k+1))...
-(.5*v/dx)*(u(k+1)-u(k-1))+Q(u(k)));
end
unext(1)=(g(n*dt)*dx-b*unext(2))/(a*dx-b);
unext(J+1)=(h(n*dt)*dx+d*unext(J))/(c*dx+d);
for k=1:J+1
u(k)=unext(k);
end
if n==counter*profiles
z=[0:L/J:L];
plot(z,u)
counter=counter+1;
end
end
function IC=f(x) %intial condition
IC=5*x^2*(1-x)^6;
function LBC=g(t) %left boundary condition
LBC=0;
function RBC=h(t) %right boundary condition
RBC=0;
function rate=Q(s) %reaction rate
rate=0;
```

Implicit Method

The preceding method is called *explicit* because it permits the direct calculation of concentrations at the next time step from concentrations at the preceding
time step. On the other hand, implicit methods can be developed where at each
new time step a system of equations must be solved to determine the concentrations at that time. The payoff is that implicit schemes have better stability

properties than explicit schemes and the time step is not severely restricted, as required by the stability condition (A.9).

A common family of implicit schemes is defined by the **Crank–Nicolson method**. It provides for a weighted average of the spatial derivatives at the nth and the $n + 1$st time levels. Therefore, instead of (A.4), we have

$$\frac{U_j^{n+1} - U_j^n}{\Delta t} = \lambda \left(D \frac{U_{j-1}^n - 2U_j^n + U_{j+1}^n}{\Delta x^2} - v \frac{U_{j+1}^n - U_{j-1}^n}{2\Delta x} \right)$$
$$+ (1 - \lambda) \left(D \frac{U_{j-1}^{n+1} - 2U_j^{n+1} + U_{j+1}^{n+1}}{\Delta x^2} - v \frac{U_{j+1}^{n+1} - U_{j-1}^{n+1}}{2\Delta x} \right)$$
$$+ F(U_j^n)$$

where λ is a weight factor between zero and one. If $\lambda = 1$, then we obtain the explicit scheme (A.5) where we can compute the U_j^{n+1} directly. Otherwise, if $\lambda \neq 1$ we obtain

$$aU_{j-1}^{n+1} + bU_j^{n+1} + cU_{j+1}^{n+1} = \lambda (r + s) U_{j-1}^n + (1 - 2\lambda r) U_j^n \quad \text{(A.10)}$$
$$+ \lambda (r - s) U_{j+1}^n + \Delta t F(U_j^n),$$

for $j = 1, ..., J - 1$, where a, b, and c are given by

$$c = -(1 - \lambda)(r - s), \quad a = 1 + 2(1 - \lambda)r, \quad b = -(1 - \lambda)(r + s).$$

The expressions (A.10) represent a system of $J - 1$ equations in the $J - 1$ unknowns $U_1^{n+1}, ..., U_{J-1}^{n+1}$, representing the concentrations at the $n + 1$st time step. Observe that the right-hand side of equation (A.10) is fully determined by the known values of the concentration at the previous time step. The system of equations (A.10) is a tridiagonal system of the form

$$\begin{pmatrix} a & c & & & & & \\ b & a & c & & & & \\ & b & a & c & & & \\ & & b & a & c & & \\ & & & \cdot & \cdot & \cdot & \\ & & & & \cdot & \cdot & \cdot \\ & & & & b & a & c \\ & & & & & b & a \end{pmatrix} \begin{pmatrix} U_1^{n+1} \\ U_2^{n+1} \\ U_3^{n+1} \\ \cdot \\ \cdot \\ \cdot \\ U_{J-2}^{n+1} \\ U_{J-1}^{n+1} \end{pmatrix} = \begin{pmatrix} B_1^n \\ B_2^n \\ B_3^n \\ \cdot \\ \cdot \\ \cdot \\ B_{J-2}^n \\ B_{J-1}^n \end{pmatrix}$$

where the B_j^n are the right-hand sides of the difference equations given by

$$B_j^n = \lambda (r + s) U_{j-1}^n + (1 - 2\lambda r) U_j^n + \lambda (r - s) U_{j+1}^n + \Delta t F(U_j^n).$$

The triadiagonal system can be solved easily by forward-and-back substitution.

There are, of course, a large number of books, monographs, and papers on FDM. A good introductory text is Morton and Mayers (1994).

B The Method of Lines

The method of lines (MOL) is a hybrid FDM that takes advantage of highly efficient and accurate packages for solving systems of ordinary differential equations. The idea is to discretize only in space and use an ODE package to solve the resulting ordinary differential equations in time.

Let us consider the advection–dispersion–adsorption problem from the last section, equations (A.1)–(A.3).

The first step is to discretize the spatial interval $[0, L]$ by defining the nodes

$$x_j = j\Delta x, \quad j = 0, 1, \ldots, J; \quad \Delta x = \frac{L}{J}.$$

We do not discretize in time, and we denote the approximation of the solution $u(x_j, t)$ at the jth node by $U_j = U_j(t)$. The function $u(x_j, t)$ describes the time evolution of the concentration at the location x_j, and $U_j(t)$ is its approximation. The PDE (A.1) can be approximated at each jth node by

$$\frac{dU_j}{dt} = D\frac{U_{j-1} - 2U_j + U_{j+1}}{\Delta x^2} - v\frac{U_{j+1} - U_{j+1}}{2\Delta x} + F(U_j), \quad j = 1, 2, \ldots J - 1.$$
(B.1)

This is a system of $J - 1$ ordinary differential equations for the $J - 1$ unknown functions $U_1(t), U_2(t), \ldots, U_{J-1}(t)$. Each equation relates the concentrations along three adjacent spatial lines. The functions $U_0(t)$ and $U_J(t)$, both required in (B.1), are given by the boundary conditions

$$U_0(t) = g(t), \quad U_J(t) = h(t),$$

and the initial conditions are given by

$$U_j(0) = f(j\Delta x), \quad j = 1, 2, \ldots, J - 1.$$

The problem is now set up for solution by an ODE solver, like a Runge–Kutta method.

The following MATLAB program calls the package **ode45**, a medium-order method for nonstiff problems, to solve the following advection–dispersion problem:

$$u_t = 0.8u_{xx} - 4u_x - \frac{2u}{u + 1}, \quad 0 < x < 1, \ t > 0, \tag{B.2}$$

$$u(0, t) = 0, \quad u(1, t) = 0, \quad t > 0, \tag{B.3}$$

$$u(x, 0) = 6x(1 - x)^6, \quad 0 < x < 1. \tag{B.4}$$

The notation is the same as in the previous program. The solution (concentration) surface is shown in figure 3.

MATLAB Program (MOL)

```
function concenMOL( )
global J
global L
L=1.0; J=30; dx=L/J;
for j=1:J+1
u0(j)=f((j-1)*dx);
end
init=u0;
[T,U] = ode45('conMOL',[0:0.001:.03],init);
plot(U(:,:)');
figure(2)
Xvals=[0:dx:L];
[X,Y] = meshgrid(Xvals,T);
mesh(Y',X',U')
title('Advection-Dispersion-Adsorption Equation')
zlabel('Conc')
colormap(hsv)
axis ij
xlabel('time')
ylabel('length')
view(-40,30)
function initial=f(x)
initial=6*x*(1-x)^6;

function du = conMOL(t,u)
global L;
global J;
dx=L/J;
du = zeros(J+1,1); % a column vector
u(1)=0;u(J+1)=0;
for j=2:J
du(j) = 0.6*(1/dx^2)*(u(j-1)-2*u(j)+u(j+1))-8*(0.5/dx)
*(u(j+1)-u(j-1))-2*u(j)/(u(j)+1);
end
```

Several tutorial books are available that illustrate MATLAB programming techniques for a variety of engineering and scientific problems. The manuals packaged with the software (e.g., *Using MATLAB* (1998)) provide a good introduction to the language with several examples of solving differential equations.

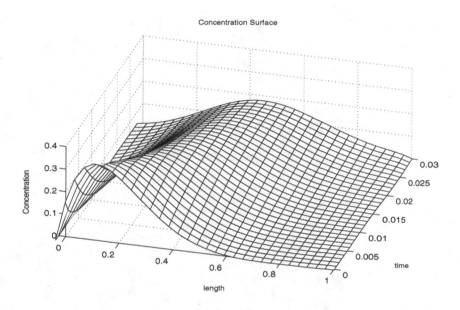

Figure 3: Concentration surface for the initial boundary value problem (B.2)–(B.4).

C Numerical Inversion of Transforms

Integral transforms are an important tool in quantitative hydrogeology, and the Laplace transform is chief among them. Although many problems lend themselves to an application of Laplace transform methods, the inversion problem back from the transform domain is often impossible to perform analytically, and therefore numerical methods of inversion are required. In this appendix we briefly discuss two numerical methods, the Talbot algorithm and the Stehfest algorithm. These are commonly used methods in hydrogeology. It is assumed that the reader has a basic knowledge of transform methods. The text by Churchill (1958) still stands as an outstanding introduction to analytic, integral transform methods and their applications to partial differential equations.

The Laplace transform $F(s)$ of a function $f(t)$ is defined by the formula

$$F(s) = \int_0^\infty f(t)e^{-st}dt,$$

and the inversion formula is

$$f(t) = \frac{1}{2\pi i} \int_B F(s)e^{st}ds, \tag{C.1}$$

where B is a Bromwich path in the complex plane given by an infinite vertical line from $a - i\infty$ to $a + i\infty$. Here, a is chosen so that B lies to the right of all the singularities (poles, essential singularities) of $F(s)$.

Note that the inversion formula (C.1) involving the Bromwich path B is difficult to compute numerically because the factor $e^{st} = e^{t\operatorname{Re} s}e^{it\operatorname{Im} s}$ in the integrand oscillates rapidly for large $\operatorname{Im} s$. The idea of the Talbot algorithm is to transform the inversion integral (C.1) to one over a closed, bounded real interval and then use a numerical quadrature formula like the trapezoid rule. The inversion formula is equivalent to

$$f(t) = \frac{1}{2\pi i} \int_L F(s)e^{st}ds,$$

where L is a curve beginning and ending in the left half plane and encloses the singularities of F. See figure 4.

205

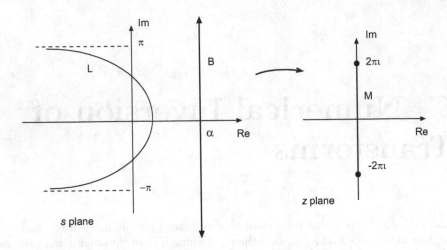

Figure 4: The complex s-plane showing the Bromwich path B and the the deformed path L, and thecomplex z-plane showing the interval M. The region outside L is mapped the the right half-plane.

We then map L onto a closed interval $M = [-2\pi i, 2\pi i]$ on the imaginary axis via the transformation

$$s = S(z) = \frac{z}{1 - e^{-z}}.$$

Then

$$
\begin{aligned}
f(t) &= \frac{1}{2\pi i} \int_M F(S(z)) e^{S(z)t} S'(z) dz \\
&= \frac{1}{2\pi i} \int_M Q(z) dz,
\end{aligned}
$$

where

$$Q(z) = F(S(z)) e^{S(z)t} S'(z).$$

Here, $S'(z) = (1 - (1 + z)e^{-z})/(1 - e^{-z})^2$. Now rotate the last integral by ninety degrees to obtain the real interval of integration $[-2\pi, 2\pi]$ and then use the trapezoidal rule. That is, let $w = -iz$. Then

$$f(t) = \frac{1}{2\pi} \int_{-2\pi}^{2\pi} Q(iw) dw.$$

The trapezoidal rule then gives (with n odd)

$$f(t) \cong \frac{1}{n} \left[Q(2\pi i) + Q(-2\pi i) + 2 \sum_{j=1}^{n-1} Q(iw_j) \right],$$

where

$$w_j = 2\pi(\frac{2j}{n} - 1).$$

Observe that $Q(2\pi i) = Q(-2\pi i) = 0$.

As an example, the following Maple program inverts the transform

$$F(s) = \frac{1}{(1+s)^5}.$$

The exact inverse transform is

$$f(t) = \frac{1}{24}t^4 e^{-t},$$

and the numerical inversion is highly accurate; for $0 \leq t \leq 13$ the plots of the exact inverse and the approximation are indistinguishable.

Maple Program (Talbot Algorithm)

```
restart:
n:=25: Digits:=16:
S:=z->z/(1-exp(-z)): Sprime:=z->(1-(1+z)*exp(-z))/(1-exp(-z))^2:
F:=s->1/(1+s)^5:
Q:=(t,z)->evalf(exp(S(z)*t)*F(S(z))*Sprime(z)):
w:=seq(evalf(2*Pi*(2*j/n-1)),j=1..n):
f:=t->Re((2/n)*sum(Q(t,I*w[j]),j=1..n-1)):
plot({f(t),t^4*exp(-t)/24},t=0..13);
```

The **Stehfest algorithm** relies on defining a special sequence of weights V_j, $j = 1, ..., n$, (n even) and approximating the inversion integral (C.1) by the sum

$$f(t) \cong \frac{\ln 2}{t} \sum_{j=1}^{n} V_j F(\frac{j \ln 2}{t}).$$

The weights are defined by

$$V_j = (1)^{j+\frac{n}{2}} \sum_{k=[\frac{j+1}{2}]}^{\min(j,n/2)} \frac{k^{n/2}(2k)!}{k!(k-1)!(j-k)!(2k-j)!(\frac{n}{2}-k)!}.$$

The reader might compare the Stehfest inversion formula to the Post–Widder formula

$$f(t) \cong \lim_{k\to\infty} \left(\frac{(-1)^k}{k!} F^{(k)} \left(\frac{k}{t}\right) \left(\frac{k}{t}\right)^{k+1} \right).$$

As an example we use the Stehfest algorithm programmed in Maple to invert the transform

$$F(s) = \frac{1}{\sqrt{1+s^2}}.$$

The exact inverse is the Bessel function $J_0(t)$, and the Stehfest algorithm performs well.

Maple Program (Stehfest Algorithm)

```
Restart: n:=20:
F:=s->1/sqrt(1+s^2):
> V:=seq((-1)^(j+n/2)*sum(k^(n/2)*(2*k)!/((n/2-k)!*k!*(k-1)!
    *(j-k)!*(2*k-j)!),k=iquo(j+1,2)..min(j,n/2)),j=1..n):
> f:=t->(ln(2)/t)*sum(V[j]*F(j*ln(2)/t),j=1..n):
> plot({f(t),BesselJ(0,t)},t=0..8);
```

Of course, the issue is when to use what algorithm. This is a difficult question and different algorithms may be appropriate for different functions, depending upon their behavior at infinity; some perform well for different ranges of time t. The lack of universal methods for inversion stems from the fact that the space of Laplace transformable functions is too large. The articles listed in the references [Ang et al. (1989), Davies and Martin (1979), De Hoog et al. (1982), Stehfest (1970), Talbot (1979)] should give the reader a good perspective of the issues involved.

D Notation and Symbols

The list below gives some of the standard quantities in hydrogeology and the symbols used in this monograph to represent those quantities. Some symbols are used to denote different quantities, but the context should make it clear which quantity is under discussion.

average velocity (Darcy velocity divided by porosity), v

concentration of solutes, C
concentration of solutes (dimensionless or generic), u
concentration of sorbed particles, S, s

Darcy velocity or filtration velocity, V
dispersion constant, D
dispersivity, α
density of water or fluid, ρ

flux or discharge, Q

head or hydraulic head, h
hydraulic conductivity, K
hydrodynamic dispersion constant, D

molecular diffusion constant, D

permeability, k
porosity, ω
pressure, p

speed of traveling waves, c

viscosity, μ

References

1. M. Abramowitz and I. Stegun (1965), *Handbook of Mathematical Functions*, Dover Publications, New York.

2. D. J. Acheson (1990), *Elementary Fluid Dynamics*, Clarendon Press, Oxford.

3. D. D. Ang, J. Lund, and F. Stenger (1989), Complex variable and regularization methods of inversion of the Laplace transform, *Math. Comp.* 53(188), 589–608.

4. J. F. G. Auchmuty, J. Chadam, E. Merino, P. Ortoleva, and E. Ripley (1986), The structure and stability of propagating redox fronts, *SIAM J. Appl. Math.* 46(4), 588–604.

5. P. Bassinini and A. R. Elcrat (1997), *Partial Differential Equations*, Plenum Press, New York.

6. J. Bear (1988), *Dynamics of Fluids in Porous Media*, Dover Publications, New York (reprint of 1972 edition published by Elsevier, New York).

7. R. A. Berner (1971), *Principles of Chemical Sedimentology*, McGraw-Hill, New York.

8. R. B. Bird, W. E. Stewart, and E. N. Lightfoot (1960), *Transport Phenomena*, John Wiley and Sons, New York.

9. W. J. P. Bosma and S. van der Zee (1993), Analytical approximation for nonlinear adsorbing solute transport and first-order approximation, *Trans. Porous Media* 11, 33–43.

10. N. F. Britton (1986), *Reaction-Diffusion Equations and Their Applications to Biology*, Academic Press, London.

11. J. R. Cannon (1984), *The One-Dimensional Heat Equation*, Addison-Wesley Pub. Co., Menlo Park, CA.

12. H. S. Carslaw and J. C Jaeger (1959), *Heat Conduction in Solids,* 2nd ed., Clarendon Press, Oxford.

13. A. J. Chorin and J. E. Marsden (1993), *A Mathematical Introduction toFluid Mechanics*, 3rd ed., Springer-Verlag, New York.

14. R.V. Churchill (1958), *Operational Mathematics*, 2nd ed, McGraw-Hill, New York.

15. S. Cohn, and J. D. Logan (1995a), Mathematical analysis of reactive-diffusive model of the dispersal of a chemical tracer with nonlinear convection, *Math. Models and Meth. in Appl. Sci.*, 5(1), 29–46.

16. S. Cohn, and J. D. Logan (1995b), Existence of solutions to equations modeling colloid transport in porous media, *Comm. on Appl. Nonlinear Anal.* 2(2), 33–44.

17. S. Cohn, J. D. Logan, and T. S. Shores (1996), Stability of traveling waves for a solute transport problem in porous media, *Canadian Appl. Math. Quart.* 4(3), 243–263.

18. S. Cohn, G. Ledder, and J. D. Logan (2000), Analysis of a filtration model in porous media, *PanAmerican Math. J.* 10(1), 1–16.

19. J. Crank (1975), *Mathematics of Diffusion*, 2nd ed.,Clarendon Press, Oxford.

20. B. Davies and B. Martin (1979), Numerical inversion of the Laplace transform: a survey and comparison of methods, *J. Comp. Physics* 33, 1–32.

21. J. I. Diaz and I. Stakgold (1995), Mathematical aspects of combustion of a solid by a distributed isothermal gas reaction, *SIAM J. Math.Anal.* 26(2), 305–328.

22. F. R. De Hoog, J. H. Knight, and A. N. Stokes (1982), An improved method for numerical inversion of Laplace transforms, *SIAM J. Sci. Stat. Comput.* 3(3), 357–366.

23. P. A. Domenico and F. W. Schwartz (1990), *Physical and Chemical Hydrogeology*, John Wiley and Sons, New York.

24. P. G. Drazin and R. S. Johnson (1989), *Solitons: An Introduction*, Cambridge University Press, Cambridge.

25. L. Dresner (1983), *Similarity Solutions of Nonlinear Partial Differential Equations*, Pitman Publishing Inc., Boston.

26. H. Engler (1985), Relations between travelling wave solutions of quasilinear parabolic equations, *Proc. Am. Math. Soc.* 93, 297–302.

27. M. S. Espedal, A. Fasano, and A. Mikelic, eds (2000), *Filtration in Porous Media and Industrial Application*, Lecture Notes in Mathematics, No. 1734, Springer-Verlag, Berlin.

28. L. Evans (1998), *Partial Differential Equations*, Amer. Math. Soc., Providence.

29. A. Fasano (editor) (2000), *Complex Flows in Industrial Processes*, Birkhauser, Boston.

30. C. W. Fetter (1993), *Contaminant Hydrogeology*, Macmillan Pub. Co., New York.

31. J. N. Flavin and S. Rionero (1996), *Qualitative Estimates for Partial Differential Equations*, CRC Press, Boca Raton.

32. A. C. Fowler (1997), *Mathematical Models in the Applied Sciences*, Cambridge University Press, Cambridge.

33. R. A. Freeze and J. A. Cherry (1979), *Groundwater*, Prentice-Hall, Englewood Cliffs, NJ.

34. A. Friedmann (1964) *Partial Differential Equations of Parabolic Type*, Prentice-Hall, Englewood Cliffs.

35. L. W. Gelhar (1993), *Stochastic Subsurface Hydrology*, Prentice-Hall, Englewood Cliffs, NJ.

36. *Geochemistry of Hydrothermal Ore Deposits, 3rd ed*, (1997), edited by H.L. Barnes, John Wiley and Sons, New York.

37. P. Grindrod (1993), Some reactive transport, dispersal and flow problems associated with geological disposal of radioactive wastes, in: *Ordinary and Partial Differential Equations IV*, (ed. B.D. Sleeman and R.J. Jarvis). Longman.

38. P. Grindrod (1996), *The Theory and Applications of Reaction-Diffusion Equations, 2nd ed.*, Oxford University Press, New York.

39. R. B. Guenther and J. W. Lee (1996), *Partial Differential Equations of Mathematical Physics and Integral Equations*, Dover Publications, New York.

40. P. Hartman (1982), *Ordinary Differential Equations*, 2nd ed., Birkhauser, Boston.

41. J. F. Hermance (1999), *A Mathematical Primer on Groundwater Flow*, Prentice-Hall, Upper Saddle River, NJ.

42. R. J. P. Herzig, D. M. Leclerc, and P. LeGoff (1970), Flow of suspension through porous media, *Ind. Eng. Chem.* 62(5), 8–35.

43. D. Henry (1981), *Geometric Theory of Semilinear Parabolic Equations*, Springer-Verlag, New York.

44. D. Hilhorst and J. Hulshof (1991), An elliptic-parabolic problem in com-
    bustion theory: convergence to traveling waves, *Nonl. Analy.* 17(6), 519–
    546.

45. D. Hilhorst and M. A. Peletier (1997), Convergence to traveling waves in a
    reaction-diffusion system arising in contaminant transport, *Nonl. Analy.*
    17(6), 519–546.

46. M. R. Homp and J. D. Logan (1997), Contaminant transport in fractured
    media with sources in the porous domain, *Transport in Porous Media* 29,
    341–353.

47. M. R. Homp and J. D. Logan (1999), Shocks and wave fronts in a convection-
    diffusion-adsorption model with bounded flux, *Comm. Appl. Nonl. Analy.*
    4(3), 1–15.

48. F. John (1981), *Partial Differential Equations*, 4th ed., Springer-Verlag,
    New York.

49. J. Kevorkian (2000), *Partial Differential Equations*, 2nd ed., Springer-
    Verlag, New York.

50. J. Kevorkian and J. Cole (1981), *Perturbation Methods in Applied Math-
    ematics*, Springer-Verlag, New York.

51. R. Knobel (2000), *An Introduction to the Mathematical Theory of Waves*,
    American Mathematical Society, Providence, RI.

52. A. Kurganov and P. Rosenau (1997), Effects of a saturating dissipation in
    Burgers-type equations, *Comm. Pure Appl. Math.* 50, 753–771.

53. A. Kurganov, D. Levy, and P. Rosenau (1998), On Burgers-type equations
    with nonmonotonic dissipative fluxes, *Comm. Pure Appl. Math.* 51, 443–
    472.

54. A. C. Lasaga (1998), *Kinetic Theory in the Earth Sciences*, Princeton
    University Press, Princeton, NJ.

55. G. Ledder and J. D. Logan (2000), Contamination and remediation waves
    in a filtration model, *Appl. Math. Lett.* 13, 75–84.

56. G. Ledder and J. D. Logan (2001), Corrigendum: Contamination and
    remediation waves in a filtration model, *Appl. Math. Lett.* in press.

57. C. C. Lin and L. A. Segel (1989), *Mathematics Applied to Deterministic
    Problems in the Natural Sciences*, reprinted by SIAM, Philadelphia.

58. P. C. Lichtner (1985), Continuum model for simultaneous chemical reac-
    tions and mass transport in hydrothermal systems, *Geochim. Cosmochim.
    Acta*, 49, 779–800.

59. B. E. Logan (1999), *Environmental Transport Processes*, Wiley-Interscience, New York.

60. J. D. Logan (1987), *Applied Mathematics: A Contemporary Approach*, Wiley-Interscience, New York.

61. J. D. Logan (1994), *Introduction to Nonlinear Partial Differential Equations*, Wiley-Interscience, New York.

62. J. D. Logan (1996), Solute transport in porous media with scale-dependent dispersion and periodic boundary conditions, *J. Hydrol.* 184, 261–276.

63. J. D. Logan (1997a), Stability of wavefronts in a variable porosity model, *Appl. Math. Lett.* 10(6), 83–89.

64. J. D. Logan (1997b), Weighted L^2 stability of traveling waves in a porous media, *Comm. Appl. Nonl. Analy.* 4(1), 55–62.

65. J. D. Logan (1997c), *Applied Mathematics: 2nd ed.*, Wiley-Interscience, New York.

66. J. D. Logan (1998a), Wave front solutions to a filtration equation with growth, *Comm. Appl. Nonl. Analy.* 5(1), 33–43.

67. J. D. Logan (1998b), *Applied Partial Differential Equations*, Springer-Verlag, New York.

68. J. D. Logan (1999), Reaction fronts in porous media with varying porosity. An exact solution, *Comm. Appl. Nonl. Analy.* 6(4), 45–50.

69. J. D. Logan (2001), Approximate wave fronts in a class of reaction-diffusion equations, *Comm. Appl. Nonl. Analy.*, in press.

70. J. D. Logan and G. Ledder (1995), Travelling waves for a nonequilibrium, two-site, nonlinear sorption model, *Appl. Math. Modelling* 19, 270–277.

71. J. D. Logan, G. Ledder, and M. Homp (1998), A singular perturbation problem in fractured media with parallel diffusion, *Math. Models and Meth. in Appl. Sci.* 8(4), 645–655.

72. J. D. Logan, M. R. Petersen, and T. S. Shores (2001), Numerical analysis of porosity-mineralogy changes in porous media, *Appl. Math. and Comp.*, in press.

73. J. D. Logan and V. Zlotnik (1995), The convection-diffusion equation with periodic boundary conditions, *App. Math. Lett.* Vol 8, No 3, 55–61.

74. J. D. Logan, V. Zlotnik, and S. Cohn, (1996), Transport in fractured porous media with time-periodic boundary conditions, *Math. Comput. Modelling* 24(9), 1–9.

75. G. de Marsily (1986), *Quantitative Hydrogeology*, Academic Press, Inc., San Diego.

76. *MATLAB, version 5* (1988), The Math Works, Inc., Natick, MA.

77. J. D. Murray (1993), *Mathematical Biology*, Springer-Verlag, New York.

78. R. McOwen (1996), *Partial Differential Equations*, Prentice-Hall, Englewood Cliffs.

79. K. W. Morton and D. F. Mayers (1994), *Numerical Solution of Partial Differential Equations*, Cambridge University Press, Cambridge.

80. R. Murray and J. X. Xin (1998), Existence of traveling waves in a biodegradation model for organic contaminants, *SIAM J. Appl. Math* 30(1), 72–94.

81. J. Ockendon, S. Howison, A. Lacey, and A. Movchan (1999), *Applied Partial Differential Equations*, Oxford Univ. Press, Oxford.

82. J. E. Odencrantz, A. J. Valocchi, and B. E. Rittman (1993), Modeling the interaction of sorption and biodegradation on transport in ground water in situ bioremediation systems, in *Proc. of 1993 Groundwater Modeling Conference*, ed. by E. Poeter, S. Ashlock and J. Proud, pp. 2-3 to 2-12, Int. Ground Water Model. Cent., Golden, CO.

83. S. Osher and J. Ralston (1982), L^1 stability of travelling waves with applications to convective porous media, *Comm. Pure Appl. Math.* 35, 737–749.

84. P. J. Ortoleva (1994), *Geochemical Self-Organization*, Clarendon Press, Oxford.

85. P. Orteleva, G. Auchmuty, J. Chadam, J. Hettmer, E. Merino, C. Moore, and E. Ripley (1986), Redox front propagation and banding modalities, *Physica* 19D, 334–354.

86. S. Oya and A. J. Valocchi (1997), Characterization of traveling waves and analytical estimation of pollutant removal in one-dimensional subsurface bioremediation modeling, *Water Resour. Res.* 33(5), 1117–1127.

87. M. A. Peletier (1997), *Problems in Degenerate Diffusion*, Dissertation, University of Leiden.

88. O. M. Phillips (1991), *Flow and Reactions in Permeable Rocks*, Cambridge University Press, Cambridge.

89. P.Ya. Polubarinova-Kochina (1962), *Theory of Groundwater Movement*, Princeton Univ. Press, Princeton.

90. M. Protter and H. Weinberger (1967), *Maximum Principles in Differential Equations*, Prentice-Hall, Englewood Cliffs.

91. R. Rajagopolan and C. Tien (1976), Deep bed filtration, *AIChE J.* 22(3), 523–533.

92. R. Rajagopolan and C. Tien (1979), The theory of deep bed filtration, in *Progress in Filtration and Separation, 1*, ed. R.J. Wakeman, Elsevier Scientific Publications, Amsterdam.

93. M. Renardy and R. C. Rogers (1993), *An Introduction to Partial Differential Equations*, Springer-Verlag, New York.

94. J. E. Saiers, G. M. Hornberger, and L. Liang (1994), First-and second-order kinetics approaches for modeling the transport of colloidal particles in porous media, *Water Resour. Res.* 30(9), 2499-2506.

95. A. A. Samarskii, S. P. Gurdyumov, V.A. Galaktionov, and A.P. Mikhailov (1995), *Blow-up in Quasilinear Parabolic Equations*, W. deGruyter, Berlin.

96. D. H. Sattinger (1976), On the stability of waves in nonlinear parabolic systems, *Adv. Math.* 22, 312–355.

97. H. Stehfest (1970), Numerical inversion of Laplace transforms, *Comm. ACM* 13(1), 47–49; Remark on numerical inversion of Laplace transforms, *Comm. ACM* 13(10), 624.

98. J. L. Schiff (1999), *The Laplace Transform*, Springer-Verlag, New York.

99. J. Smoller (1994), *Shock Waves and Reaction Diffusion Equations*, 2nd ed., Springer-Verlag, New York.

100. B. Straughan (1992), *The Energy Method, Stability, and Nonlinear Convection*, Springer-Verlag, New York.

101. W. Strauss (1993), *Elementary Partial Differential Equations*, John Wiley and Sons, New York.

102. E.A. Sudicky and E.O. Frind (1982), Contaminant transport in fractured media: Analytical solution for a system of parallel fractures, *Water Resour. Res.* 18(6), 1634–1642.

103. N.-Z. Sun (1995), *Mathematical Modeling of Groundwater Pollution*, Springer-Verlag, New York.

104. A. Talbot (1979), The accurate numerical inversion of Laplace transforms, *J. Inst. Maths. Applics.* 23, 97–120.

105. D. H. Tang, E.A. Sudicky, and E. O. Frind (1981), Contaminant transport in fractured porous media: analytical solution for a single fracture, *Water Resour. Res.* 17(3), 555–564.

106. G. I. Taylor (1953), Dispersion of soluble matter in a solvent flowing slowly through a tube, *Proc. Royal Soc.* A219, 186–203.

107. A. N. Tikhonov and A.A. Samarskii (1990), *Equations of Mathematical Physics,* reprinted by Dover Publications, New York.

108. C. C. Travis and E.L. Etnier (1981), A survey of sorption relationships for reactive solutes in soil, *J. Environ. Qual.* 10(1),8–17.

109. *Using MATLAB, ver 5.2,* (1998), The Math Works, Inc., Natick, MA.

110. S. van der Zee (1990), Analytical traveling wave solutions for transport with nonlinear and nonequilibrium adsorption, *Water Resour. Res.* 26(10), 2563–2578. Correction: *Water Resour. Res.* 27(5), 983.

111. C. J. van Duijn and J. M. de Graaf (1987), Limiting profiles in contaminant transport through porous media, *SIAM J. Math. Anal.* 18(3), 728–743.

112. C. J. van Duijn and P. Knabner (1991), Solute transport in a porous media with equilibrium and nonequilibrium multiple-site adsorption: travelling waves, *J. reine angew. Math.* 115, 1–49.

113. C. J. van Duijn and P. Knabner (1992a), Traveling waves in the transport of reactive solutes through porous media: adsorption and binary ion exchange, Part 1, *Transport in Porous Media* 8, 167–194.

114. C. J. van Duijn and P. Knabner (1992b), Traveling waves in the transport of reactive solutes through porous media: adsorption and binary ion exchange, Part 1, *Transport in Porous Media* 8, 199–225.

115. C. J. van Duijn, P. Knabner, and S. van der Zee (1993), Traveling waves during the transport of reactive solutes in porous media: Combination of Langmuir and Freundlich isotherms, *Advances in Water Resources* 16, 97–105.

116. H. van Duijn and P. Knabner (1994), Flow and reactive transport in porous media induced by well-injection: similarity solution, *IMA J. Appl. Math.* 52, 177–200.

117. M. Th. van Genuchten and W.J. Alves (1982), *Analytical Solutions of the One-Dimensional Convective-Dispersive Solute Transport Equation,* U.S. Department of Agriculture, Tech. Bulletin No. 1661.

118. A. Verruijt (1970), *Theory of Groundwater Flow,* MacMillan, New York.

119. A. I. Volpert and V. A. Volpert (1994), *Traveling Wave Solutions of Parabolic Systems,* Amer. Math. Soc., Providence, RI.

120. W. Walter (1998), *Ordinary Differential Equations,* Springer-Verlag, New York.

121. G. B. Whitham (1974), *Linear and Nonlinear Waves,* Wiley-Interscience, New York.

122. D. Widder (1970), *The Heat Equation*, Academic Press, New York.

123. W. J. Wnek, D. Gidaspow, and D. T. Wasan (1975), The role of colloid chemistry in modeling deep bed liquid filtration, *Chem. Eng. Sci.* 30, 1035–1047.

124. W. Wolesensky and J. D. Logan (2001), Nonlocal, advective filtration models, in preparation.

125. J. R. Wood (1987), A model for dolomitization by pore fluid, in: *Physics and Chemistry of Porous Media* (ed.J.R. Banavar, et al), Amer. Inst. Physics, New York.

126. J. R. Wood and T. A. Hewett (1982), Fluid convection and mass transfer in porous limestones: a theoretical model, *Geochim. Cosmochim. Acta* 46, 1707–1713.

127. H. Zhang (1992), On a nonlinear singular diffusion problem: convergence to a traveling wave, *Nonl. Analy.* 19, 1111–1120.

128. V. Zlotnik and J. D. Logan (1996), Boundary conditions for convergent radial tracer tests and effect of well bore mixing volume, *Water Resour. Res.* 32(7), 2323-2328.

Index

Interdisciplinary Applied Mathematics